KU-334-696

Setting-out
Procedures

CIRIA, the Construction Industry Research and Information Association, is an independent non-profit-distributing body which initiates and manages research and information projects on behalf of its members. CIRIA projects relate to all aspects of design, construction, management, and performance of buildings and civil engineering works. Details of other CIRIA publications, and membership subscription rates, are available from CIRIA at the address below.

Lists of all available CIRIA publications can be obtained from:
CIRIA
6 Storey's Gate
London SW1P 3AU
Tel. 01–222–8891

#16681220

Setting-out Procedures

B. M. Sadgrove MA CEng MICE

ANDERSONIAN LIBRARY
★
WITHDRAWN
FROM
LIBRARY
STOCK
★
UNIVERSITY OF STRATHCLYDE

D
624
SAD

CIRIA

**Construction
Industry
Research
and
Information
Association**

Butterworths
London Boston Durban
Singapore Sydney Toronto Wellington

UNIVERSITY OF
STRATHCLYDE LIBRARIES

All rights reserved. No part of this publication may be
reproduced or transmitted in any form or by any means,
including photocopying and recording, without the written
permission of the copyright holder, application for which should
be addressed to the Publishers. Such written permission must also
be obtained before any part of this publication is stored in a
retrieval system of any nature.

This book is sold subject to the Standard Conditions of Sale of
Net Books and may not be re-sold in the UK below the net price
given by the Publishers in their current price list.

First published 1988

© CIRIA, 1988

British Library Cataloguing in Publication Data

Sadgrove, B. M.
   Setting-out procedures
   1. Building sites
   I. Title   II. Construction Industry Research
and Information Association
   624   TH375

   ISBN 0–408–02745–2

Library of Congress Cataloging-in-Publication Data

Sadgrove, B. M. (Barry Martin)
   Setting-out procedures / B. M. Sadgrove.
      p.   cm.
   Bibliography: p.
   Includes index.
   ISBN 0–408–02745–2
   1. Surveying.   I. Construction Industry Research and
Information Association.   II. Title.
TA625.S23   1988
624——dc19   87–23053

UNIVERSITY OF STRATHCLYDE

29 JUN 1989

Photoset by Latimer Trend & Company Ltd, Plymouth, Devon
Printed and bound in England by Anchor-Brendon Ltd, Tiptree, Essex

Preface

This pocket book decribes recommended procedures for the setting-out of building and civil engineering works. These procedures are based on practical experience, have been proven over a number of years and are applicable to the majority of construction contracts.

Adoption of these procedures should reduce the incidence of errors, and the costs of putting them right. It is hoped that acceptance and use of these procedures will result in better communication and understanding between architects, consulting engineers, resident engineers and clerks of works and contractors' staff and foremen.

Recommendations are made for the marking and/or colour coding of pegs, profiles and so on. As far as practicable, these accord with proposals under consideration by the International Standards Organisation (ISO) and BSI, pending publication of standards by these bodies.

For the sake of convenience, the title 'site engineer' has been used throughout, but it is intended to apply to all site staff who are concerned with the actual process of setting-out. It is assumed that these site staff will have a good basic knowledge of surveying. (The familiar title of 'chainman' has also been used for convenience and is deemed to refer to either male or female.)

The site engineer is advised to report certain actions, errors and omissions. To whom the engineer should report is not generally stated, because the reporting procedure will vary from site to site. Where the term 'supervisory authority' is used, this means the Architect, Engineer or Supervising Officer or a nominated representative of one of these.

Specific recommendations as to the desired accuracy of setting out have been avoided. Site engineers should set out as accurately as the available equipment permits, whether for temporary or permanent works, and should assess their work by visual and other commonsense checks. Checks on calculator outputs are particularly important.

Although design details are not the Contractor's responsibility, the site engineer is also advised to check a number of such details, primarily to smooth the running of the Contract and to minimise any consequential delay.

This publication supersedes the CIRIA *Manual of setting-out procedures*, published in 1974. It is based on the same material as the *Manual*, rewritten in the light of developments in practice and rearranged to make it easier to use for reference in the field. New material has been included on setting out marine structures and tunnels, and to cover the use of electronic surveying equipment.

Acknowledgements

CIRIA and the author wish to acknowledge the help and advice given by the steering group in guiding the preparation and checking the technical detail of this publication. The steering group members were:

K. N. Montague BSc CEng MICE MIHT FGS (*Chairman*)
Brian Colquhoun & Partners
A. Aarons CEng FICE FIStructE
Building Design Partnership
A. I. L. Byers BSc CEng ARTC FICE
Balfour Beatty Ltd
P. K. Donaldson BSc CEng MICE
John Laing Construction Ltd
A. G. Hurrell FRICS
Property Services Agency
G. H. Reynolds CEng MICE MIHT
Department of Transport
C. Salts BSc CEng MICE MBIM
Oldham Metropolitan Borough
S. Tinnelly BSc CEng MICE MIHT
Buckinghamshire County Council
C. J. White BSc CEng MICE
Higgs and Hill Building Ltd
A. D. Wilkinson MInstCES
Construction Industry Training Board
B. M. Sadgrove MA CEng MICE
Consultant
A. R. McAvoy BSc CEng MICE
CIRIA

Thanks are also due to:

A. Bandle MSc MSRP, Chemical Inspectorate, HSE for advising on safety of laser equipment

W. A. Dawson CEng FICE, W. A. Dawson Ltd for advising on piling for marine structures

P. Clarke BSc CEng MICE and J. W. B. Marriott BEng CEng MICE, Oldham Metropolitan Borough for drafting the section on Tunnelling around curves.

J. F. T. Heardman, Construction Inspectorate, HSE for advising on general safety aspects

CIRIA's Research Manager for the project leading to this publication was A. R. McAvoy and the Technical Editor was D. R. Garner.

The illustrations were prepared by I. Clark of Night Owl Graphics.

Contents

Glossary

Bench mark	A reference point or mark of known level
Boning (-in)	See illustration on p 28
Boning rod	Upright and rail forming a tee, 1 to 1.5 m high. Used when **boning** between two other boning rods of equal height
Borrow pit	Pit supplying soil for construction of a road
Chainage	Distance (formerly chains, now metres) along a line from a datum point
Chainman	Assistant to setting-out (site) engineer
Chord points	Points on a curve defining a chord
Collimation, line of	Optical axis or line of sight of a telescope
Crossfall	Gradient on cross-section of road to shed water to one or both sides
Datum	Horizontal plane of assumed level from which other levels are determined
Deflection angle	Angle between the tangent and a chord at a point on a curve
Falsework	Temporary structure needed to construct permanent works
Formation level	Excavation level on which permanent works constructed
Formwork	Forms needed to constrain fluid concrete to desired shape
Ground distance	See pp 15 and 111
Intersection point	Intersection of tangents from **tangent points**
Invert	Lowest internal level, at a given cross-section, of a pipe, channel or tunnel
Kicker	Concrete 'step' say 50–100 mm high, used to locate vertical forms
Partial coordinates	Coordinates of a point relative to another on the same grid. They are the algebraic differences of the eastings and northings
Profile	See pp 28 and 29
Projection distance	See pp 15 and 111
Shoulder	Unpaved width at edge of road section
Sight rail	Horizontal or sloping rail of **profile**
Soffit	Lower surface of slab, bridge or similar structure
Summit	A high point on a road surface
Tangent point	Point defining start or finish of a curve
Temporary works	Temporary construction needed to construct permanent works
Total coordinates	Coordinates of a point referenced firstly by easting and secondly by northing relative to the origin of a grid
Traveller	See pp 28 and 29
Trench sheet	Steel sheet used in supporting trench sides
Valley	A low point on a road surface
Whole circle bearing	Angle measured clockwise from true or grid North

Safety aspects

Health and Safety at Work, Etc. Act 1974

Under the Act an employer is required to provide a written safety policy. Site engineers should ensure that they read and adhere to the policy. The Act also requires that employees (and self-employed persons) take reasonable care of themselves and others. In the context of setting out, the site engineer should be aware of hazards to the chainman (see Instructing the chainman) and vice versa. p. 11

Awareness of hazards

Site engineers and chainmen engaged in setting out may be relatively inexperienced. In addition, setting out does require concentration on observations or calculations. These factors do emphasize the need for both site engineers and chainmen to be aware of the potential hazards of the construction site. To counter these hazards:

- wear approved protective clothing
- be aware of activity near oneself and other members of the setting-out team
- walk and work steadily—never run
- take immediate action to correct dangerous practices or omissions

Some typical hazards

The list given below is not comprehensive but the site engineer will come to recognise other potential hazards:

- rough ground and nails projecting from discarded timber (wear approved safety footwear)
- falling objects or swinging suspended objects (wear 'Linesman'-type safety helmet—no brim)
- excavation plant, lorries, dumpers, traffic (especially on road or tunnel construction sites—wear high-visibility waistcoats)
- unsupported trenches (do not enter until support is complete, then use ladder)
- falls into unguarded excavations (ensure guards fixed)
- ladders damaged or not correctly fixed at top (report)
- scaffolding without toe boards or chest rails, with overhanging boards, etc. (report omissions)
- shot-fired nails (operator must be checked as competent—use protective glasses)
- projecting reinforcement or formwork ties
- falls into water (wear approved lifejackets)
- gases in shafts, sewers and tunnels (conform to Safe working in sewers and at sewage works, see Bibliography) p. 122
- compressed air working in tunnels (conform to CIRIA Report 44, see Bibliography) p. 122
- lasers (see Precautions under Lasers) p. 39
- falls from heights (if appropriate, use safety harness/belt)

Type	Options	Remarks
Theodolite	optical scale reading	choice of reading discrimination
	electronic scale reading	can output reading to data storage device or data processor
	laser eyepiece	indicates line of sight
	add-on EDM equipment	measures direction and distance
Total station	—	combination of theodolite, EDM equipment and data processor, gives x, y and z coordinates directly
Electromagnetic distance-measurement (EDM) equipment	manual correction	distance manually corrected for slope and ambient conditions
	automatic correction	automatically detects and corrects for slope, automatic correction for ambient conditions after setting
	combined with theodolite	*see* Theodolite, above
Optical level	non-tilting	suitable construction purposes requiring short sights only
	tilting	suitable most construction purposes
	automatic	suitable most construction purposes, but more susceptible to vibration
	precise	used with invar staff for precision levelling only
	add-on parallel plate micrometer	converts tilting or automatic level to precise level
Optical plumbing instrument (for plumbing up or down)	standard	take mean of readings in four quadrants
	automatic	single reading usually sufficient
Lasers	alignment	set by some other means to define line
	rotating (horizontal)	defines horizontal plane
	rotating (general)	defines any set plane
Optical square	single/double prism	ideal for setting out right angles for short offset distances
Gyro theodolite	—	for determining bearings relative to true North, especially underground

Surveying instruments

Assumed experience

It is assumed that the site engineer will have been instructed at least how to use a simple theodolite and an optical level. It is not practicable to cover the whole range of instruments now available or known to be under development. The engineer must refer to textbooks (*see* Bibliography) or manufacturers' literature for details of principles and operation of specific models.

Instruments that may be encountered

A table of some of the instruments that may be encountered by the site engineer is given opposite. This is not a comprehensive list but indicates the present range. Although setting out is facilitated by the modern instruments, their potential high accuracy should not be taken for granted. Furthermore site engineers should appreciate that accurate setting-out can still be achieved with the less sophisticated instruments, provided due care is taken.

Use of instruments

The manufacturer's instructions must be studied and followed. Inexperienced site engineers should also take every opportunity to work with more experienced engineers.

Keystroke Sequences

Calculator make:

Model:

Computation: Computation:

Keystroke	Display	Keystroke	Display

Checks: physical and calculation

The site engineer must acquire or develop methods of checking and cross-checking all setting-out operations, until they become second nature. For particularly critical setting out, seek an independent check by someone else. Where this is not practicable, self-checking is essential.

When principal lines and levels have been set out, the Contractor should advise the Architect, Resident Engineer or Supervising Officer as appropriate. The relevant person may arrange an independent check of these lines and levels, but the site engineer should not count on this. The responsibility for the setting out will still remain with the Contractor (see Appendix A).

p. 107

Physical checks

Where possible, make physical checks such as:
- check visually that lines and levels tie in with existing features
- check distances to nearest metre by rough taping or pacing
- check levels by 'eyeing-in' on known levels
- check that supposed right angles look to be correct
- check that falls are in the right direction
- check verticality approximately with spirit level.

Calculation checks

Calculations must *always* be checked. Most site engineers use calculators and tend to rely on them implicitly. It is, however, all too easy to input the wrong figures or press the wrong function button so that a check or independent calculation is imperative. Where possible, data for check calculations should be input in a different order to minimise the risk of miskeying an incorrect value twice. For example, if adding a column of figures, input from the top for the first total and then from the bottom as a check.

Conversions from degrees, minutes and seconds to decimal degrees are helpful for various computations but must be carefully executed.

The use of calculators for converting from rectangular to polar coordinates (and vice versa) is discussed in Appendix D. Space is provided opposite to record the precise keystrokes required for a specific model of calculator for these or other computations.

p. 113

Date ....   Levels taken for ....

From ....   To ....

Back-sight	Inter-sight	Fore-sight	Rise	Fall	Reduced level	Distance	Remarks

LEVEL BOOK HEADINGS—'RISE-AND-FALL'

Back-sight	Inter-sight	Fore-sight	Height of collimation	Reduced level	Distance	Remarks

LEVEL BOOK HEADINGS—'LINE OF COLLIMATION'

BLOGGS CONTRACTORS LTD.   No. 00001
Old Wharf
Newtown

SITE INFORMATION SHEET

Copies to: .. General Foreman
                                               .. Ganger
Date .... Site Office (Engs)
Contract .... Site Office (Q.S.)
Compiled by .... Resident Engineer

Recording and informing

Keeping records

Good records are essential for:

- accurate construction
- ready interchange of setting-out information
- dissemination of correct and unique data
- accurate measurement of completed work
- settlement of disputes with supervisory authority

Records to be maintained during the job and retained until all contractual obligations fulfilled include:

- all drawings issued by the supervising authority
- site instructions issued by the supervising authority
- coordinates of main setting-out points
- locations and levels of relevant bench marks
- original ground levels
- records of located/relocated underground services
- sketches of obstructions not shown on original drawings
- class of subsoil encountered and obstacles to excavation
- printouts and calculations for setting-out purposes
- all field books used for surveying and setting out
- copies of site information sheets issued (*see* below)
- scale drawings of the works 'as constructed'.

It is prudent to duplicate key records, especially those relating to measurement, and to store these duplicate records separately off the site.

Field documents

Site engineers should be supplied with:

- level books ruled for 'rise and fall' or 'line of collimation'
- observation books for theodolite, EDM and similar readings
- pads of site information sheets (*see* below), preferably numbered sequentially
- pads of sewer information sheets, as appropriate. *p. 53*

Site information sheets

The site engineer should use site information sheets to confirm and supplement oral information and instructions given to foremen and gangers. Use sketches where possible. Typical details to be shown include:

- essential dimensions
- offset distances from line pegs
- levels of offset level pegs
- levels of profiles and lengths of travellers
- spoil heaps, borrow pits and haul roads.

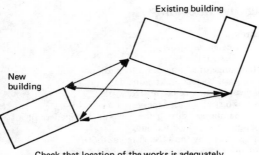

Check that location of the works is adequately shown on the drawings

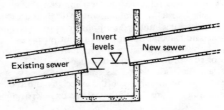

Check that existing and new levels are compatible

Temporary features can obstruct setting-out!

Initial actions

Before starting to set out the Works, do the following:

1. Review the drawings to check that:
   - the drawings are the latest issues
   - all essential dimensions are given (do not scale)
   - intermediate dimensions agree with overall dimensions
   - the location of the Works, in relation to permanent or temporary reference points, is adequately shown on the drawings
   - the level of the Works in relation to permanent features or temporary bench marks is also shown
   - critical dimensions between related components are clearly indicated

2. Walk over the site, checking:
   - that boundaries are well-defined and are as indicated on the drawings
   - that all permanent and temporary reference points and bench marks are as indicated on the drawings
   - that all visible permanent features are correctly indicated on the drawings
   - whether any permanent or temporary features may interfere with setting-out or construction of the Works
   - for evidence of hidden features that might affect setting-out or construction of the works (the Contract documents may include a warning)

3. Report any discrepancies found by the above checks *in writing immediately*. (It may take some time to rectify discrepancies)

4. Confirm any oral instructions in writing

5. Set up a system of recording and communicating information

6. Start to train the setting-out team   *p. 11*

7. Set up and/or prove setting-out stations   *p. 17*

8. Set up and/or prove bench marks   *p. 25*

9. Record and agree existing site levels and features

10. Check that the Works will tie in with existing works

11. Plan sequence of setting-out and how dimensions will be controlled

12. Do your thinking in the office!

Warning

Errors in setting-out must be reported as soon as they are discovered. Early action to correct errors will save money in the long run.

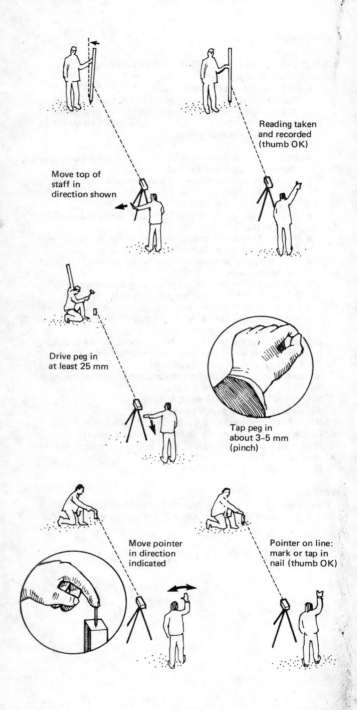

Move top of
staff in
direction shown

Reading taken
and recorded
(thumb OK)

Drive peg in
at least 25 mm

Tap peg in
about 3-5 mm
(pinch)

Move pointer
in direction
indicated

Pointer on line:
mark or tap in
nail (thumb OK)

Instructing the chainman

Basic training

A chainman is an essential member of the setting-out 'team' and a good chainman contributes considerably towards the speed and accuracy of setting out. The site engineer frequently only has the services of an inexperienced person and it is a good investment to devote time to training that person before attempting any critical setting out.

On small sites, the amount of setting out may not justify a full-time chainman and the site engineer has to 'borrow' someone. In this case, try to use the same person each time.

Check on the experience of the chainman. Where necessary, explain the basic principles of setting out, stressing the importance of accuracy and the role of the chainman in achieving accuracy and speed of setting out.

Topics that may need to be covered include the following:

- safety aspects *p. 1*
- what the various surveying instruments do *p. 2/3*
- constructing and protecting setting-out stations *p. 16/17*
- measuring with tapes *p. 20/21*
- constructing and checking temporary bench marks (TBMs) *p. 24/25*
- use of level staff and setting a peg to level *p. 26/27*
- setting up and using profiles *p. 28/29*
- using plumb bob or optical plumbing instrument *p. 30/31*
- care of equipment
- maintaining stocks of pegs, nails, paint, etc.

Action: To supplement oral instructions, provide chainman with photocopies of the pages referenced above.

Signals and commands

Agree a system of signals and commands—some suggestions are given opposite (use exaggerated, clear actions)

On difficult sites, two-way radio may be helpful.

Responsibility for chainman

The site engineer should be aware of hazards to the chainman. Ensure that the chainman is aware of these potential hazards and is wearing the correct protective clothing. *p. 1*

Chainmen have been killed by passing plant or traffic. Before signalling the chainman to move, check that it is safe to do so.

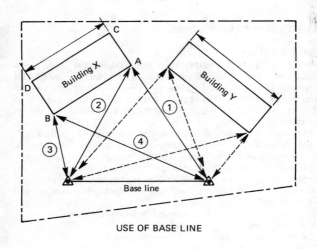

USE OF BASE LINE

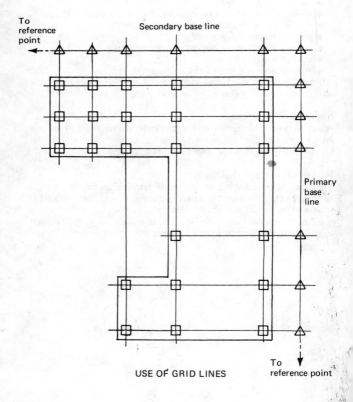

USE OF GRID LINES

Base and grid lines

Although grid coordinates are increasingly being adopted for *p. 15* setting out buildings, base lines and (setting-out) grid lines are still commonly used. If this is the case, the contract drawings should indicate the base line or grid lines to be used. A base line is suitable for a small site where reference back to the base line will be possible for all the works. Grid lines are preferable for a large site where the base line will be obscured by the works as they progress or where reference would be cumbersome. Whichever is used, the site engineer should prove any setting-out stations already provided or *p. 115* construct the stations and agree these with the supervising auth- *p. 17* ority.

Use of base line

A base line comprises two setting-out stations a given distance apart. In the example shown, point *A* of building X is set out by taping dimensions 1 and 2 from the base line and point *B* by taping dimensions 3 and 4.

The dimension *AB* is then checked against that required. Provided there is no anomaly, the remainder of the building can be set out from *AB* which effectively becomes the base-line for the building.

Corner profiles can then be set out for the building.

Use of grid lines

Usually a (primary) base line is used to set up a secondary base line and a number of setting-out stations defining the required grid lines. The example shows a rectangular grid which can conveniently coincide with base and column centres. The grid lines need not be at uniform intervals.

Where practicable, the base lines should be referenced to points well beyond the stations shown. Targets (*see* front end-papers) on existing buildings are ideal reference points.

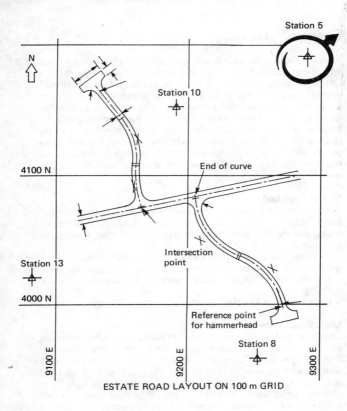

ESTATE ROAD LAYOUT ON 100 m GRID

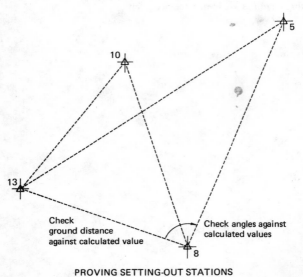

Check ground distance against calculated value

Check angles against calculated values

PROVING SETTING-OUT STATIONS

Grid coordinates

The ease of providing bearing and distance (or coordinates directly) from a theodolite and EDM equipment combined, or a total station, has increased the use of coordinates for setting out. The grid for the coordinates may be set up specifically for the site or the National Grid may be used. The general principles will be the same but note that the calculated distance between two National Grid coordinates is the *projection* distance. A correction factor must be applied to give the true horizontal or *ground* distance (*See* Appendix C).

p. 111

Whichever grid is used, setting-out stations will be provided or will have to be constructed but these need not necessarily lie on the main grid lines. The estate road layout (opposite) illustrates this.

Grid coordinates by bearing and distance
Before starting setting out, walk over the site and inspect the setting-out stations for signs of damage or displacement. Report any suspect stations in writing. Assuming that stations appear satisfactory, 'prove' them as follows:
- calculate *ground* distances between stations from coordinates
- check *ground* distances on site (correct measured distance for slope, as necessary)
- assume that two stations the correct distance apart define a base line
- calculate angles between this base line and the remaining stations
- check these angles
- report *in writing* any discrepancies outside specified limits
- agree discrepancies and actions with supervising authority

The distances and relative bearings of points to be set out should next be calculated (*see* Appendix E) and then applied on the site, correcting distance for slope as above.

p. 115

Note: Do not rely on a single calculation. Repeat calculation, using different order of input if possible, or ask somebody else to check the calculation.

Grid coordinates by total stations
Coordinates are readily set out directly with a theodolite with add-on EDM equipment or with a total station. Before defining new points, check equipment is functioning correctly by checking points of known coordinates.

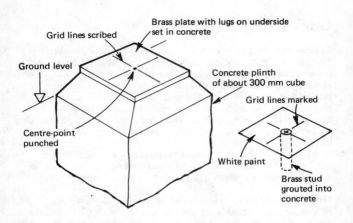

Grid lines scribed

Ground level

Brass plate with lugs on underside set in concrete

Concrete plinth of about 300 mm cube

Grid lines marked

Centre-point punched

White paint

Brass stud grouted into concrete

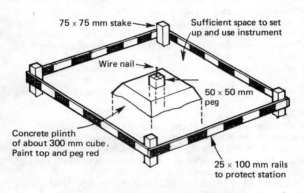

75 × 75 mm stake

Sufficient space to set up and use instrument

Wire nail

50 × 50 mm peg

Concrete plinth of about 300 mm cube. Paint top and peg red

25 × 100 mm rails to protect station

PRIMARY SETTING-OUT STATIONS

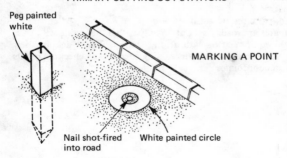

Peg painted white

MARKING A POINT

Nail shot-fired into road

White painted circle

Lines

Setting-out stations

Whether using coordinates, a base line or grid lines, setting-out stations will be required. It may be necessary to construct these stations if there is no suitable permanent hard surface on which to mark the setting-out points. Whichever is used, the setting-out stations must be sited clear of all permanent and temporary works and protected against possible damage or disturbance. Depending on the accuracy required, the stations may be as sophisticated as an Ordnance Survey triangulation pillar or as simple as a nail shot-fired into a blacktop road surface. Ensure that a theodolite may conveniently and safely be set up over each station and that there is sufficient room for the site engineer to move around the instrument.

Marking a line

When setting out a new line with a theodolite:
p. 23
- turn through required angle on both faces to make two temporary marks on indicated new line
- take mean of marks as being on new line, subject to marks being within acceptable tolerance
- make permanent mark, delete temporary marks.

The new line may be defined by one or more points (in addition to the original base line station) marked according to the need for accuracy and relative permanence. Some options are illustrated. A 'target' on existing works can be particularly valuable (*see* front end-papers). Proprietary survey markers are also available.

Indicating a line

Ranging rods are commonly used to indicate a line temporarily, e.g. for earthworks, where precision is not critical. On no account should timber battens be used for ranging purposes.

Offset pegs

In many cases, points and lines set out from the setting-out stations will be 'lost' as the permanent works are constructed. Offset pegs are necessary to allow the original points and lines to be redefined. *p. 19*

Redundant markers

Obsolete or disturbed setting-out stations, pegs or other markers should be removed to reduce risk of setting-out error and to avoid danger to people and animals when the site is vacated by the contractor.

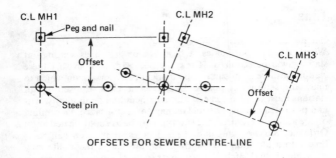

OFFSETS FOR SEWER CENTRE-LINE

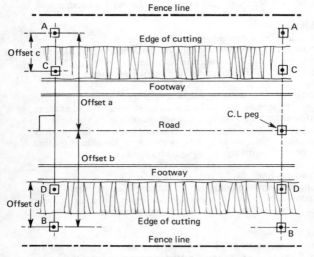

OFFSETS FOR ROAD

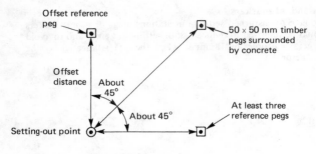

OFFSET REFERENCE PEGS

Offset pegs

Need for offset pegs
Original setting-out lines (e.g. centre-lines) and setting-out points (e.g. the centre of a circular structure) are frequently 'lost' or obscured as the permanent works are constructed. Offset points allow the original lines and points to be redefined.

Work out where offset pegs will be needed in good time and install them as soon as practicable.

Offset distances
Use of a short offset distance on a slope will minimise any error in correcting for the slope. In general, keep all offset distances short, consistent with the pegs being reasonably secure against disturbance.

All offset pegs must be located within the site boundary.

Offsets for sewer centre line
A replacement sewer is set out along a road with steel pins marking the centres of the manholes. Additional setting-out points on the line are added. Offset pegs are positioned a given perpendicular distance from the centre-line in a grass verge where they are unlikely to be disturbed during the trenching operations. A simple method of offsetting at a right angle is adequate. As far as practicable a *p. 23* constant offset distance should be used. Details of the original setting-out points, offset pegs and offset distances should be recorded on a sewer information sheet. *p. 53*

Offsets for road
The centre-line pegs are first set out. Offset pegs A and B are positioned beyond the edge of any earthworks required but within the fence lines. The offset distances *a* and *b* may not be equal if the earthworks are not symmetrical. Profiles for the earthworks can be set out and/or maintained from these offset pegs.

When the earthworks are complete offset pegs C and D can be positioned clear of the outside line of the footway, by using offsets *c* and *d* from A and B respectively. Offset pegs C and D are used to set out the kerb line.

Offset reference pegs
At least three offset pegs should be used to insure against disturbance of a critical setting-out point.

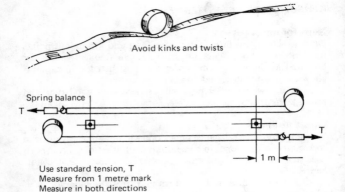

Avoid kinks and twists

Use standard tension, T
Measure from 1 metre mark
Measure in both directions

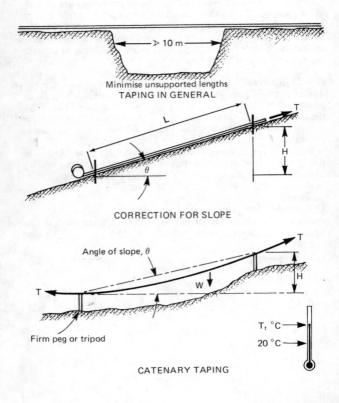

Minimise unsupported lengths
TAPING IN GENERAL

CORRECTION FOR SLOPE

CATENARY TAPING

Measurement with tapes

Choice of tapes

- Use steel tapes to BS4484 for setting out accurately
- Use plastics or steel reinforced plastics tapes only for approximate measurements
- Use invar bands for special precision, eg in tunnels p. 103

Note: Set aside and label a steel tape to BS4484 as a 'standard tape' to be used only for checking other (working) tapes

Care of tapes

- clean steel (or plastics) tape before winding into case
- clean and lightly oil steel tape at end of working day
- check (working) tapes against 'standard tape' each week and after repair.

General measuring with steel tape

- ensure steel tape is not kinked or twisted
- where possible, measure over level ground, avoiding stones, tree roots etc.
- measure close to ground to minimise wind disturbance
- apply standard tension marked on tape, using tape tension handle (*see* BRE Digest 234, Bibliography) p. 122
- avoid unsupported spans exceeding 10 m (unless true catenary)
- check overall length against sum of intermediate lengths
- take mean of two measurements in opposite directions
- correct for slope, temperature, sag
- measure from 1m mark and adjust reading accordingly

Correction for slope

Distance along slope = L (m)
Difference in level = H (m)

For slopes up to 1 in 8, horizontal distance $\simeq L - \dfrac{H^2}{2L}$ (m)

(error: not more than 3 mm in 100 m).

For steeper slopes, horizontal distance = $\sqrt{(L^2 - H^2)}$

Correction for temperature

Reading on tape = L (m)
Temperature of tape = T_t °C
Calibration temperature = 20°C
Corrected reading = $L + 11 \times 10^{-6} L(T_t - 20)$ (m)
(e.g. correction = 11 mm over 100 m for 10°C difference)

Correction for sag (catenary taping)

Measured length = L (m)
Angle of slope between supports = θ
Weight/unit length of tape = W (kg/m)
Tension applied to tape = T (kgf) (ideally standard as above)
Corrected length = $L - \dfrac{W^2 L^3 \cos^2 \theta}{24 T^2}$ (m)

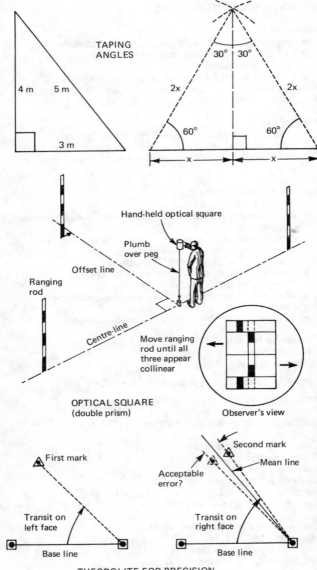

TAPING ANGLES

OPTICAL SQUARE
(double prism)

Move ranging rod until all three appear collinear

Observer's view

THEODOLITE FOR PRECISION

Horizontal angles

Simple techniques
Simple techniques are adequate for:
- setting out offset pegs from centre-lines *p. 19*
- indicating approximate limits of excavation
- generally, where accuracy is not critical

Such techniques include use of tapes, set square and optical square.

Using tapes
This technique can be used for angles of 30°, 60° and 90° but does require:
- a reasonably level site
- generally, two tapes and extra pegs

Using a set square
A site-made set square can be handy for squaring up for small excavations or setting out offset pegs fairly close to a centre-line.

Using an optical square
A hand-held optical square is an under-rated but convenient instrument for setting out offset pegs where an error of, say, 1 degree has no significant effect on the offset distance. A suitable sighting distance is 15–20 m. The principle is indicated by the illustration.

Using a theodolite
For accurate setting out, use a theodolite. Remember:
- transit with both face left and face right, using different zeros
- check, through chainman, that error is acceptable to allow mean to be taken
- ensure mean point is clearly marked.

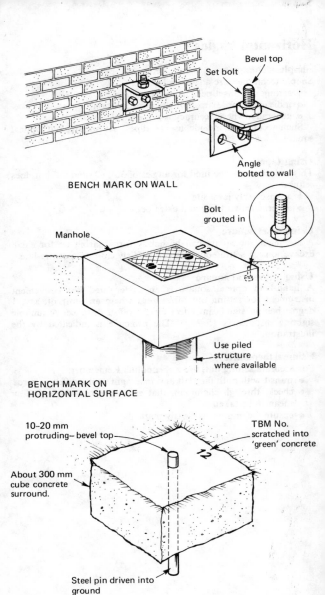

BENCH MARK ON WALL

Bevel top
Set bolt
Angle bolted to wall

Manhole
Bolt grouted in
Use piled structure where available

BENCH MARK ON HORIZONTAL SURFACE

10–20 mm protruding– bevel top
TBM No. scratched into 'green' concrete
About 300 mm cube concrete surround.
Steel pin driven into ground

TEMPORARY BENCH MARK IN FIRM GROUND

Bench marks

Each site should have a primary bench mark which may be:
- a convenient nearby Ordnance Bench Mark
- referenced to an Ordnance Bench Mark
- referenced to a given level on existing works

The supervising authority should specify which is to apply. If an Ordnance Bench Mark is to be used but which one has not been specified, obtain the current local list of Ordnance Bench Marks (see Appendix C) and agree which to use with the the the supervising authority. *p. 111*

In addition to the primary bench mark, it will be necessary to set up secondary bench marks unless all levelling can conveniently be referred to the primary bench mark. Secondary bench marks are commonly called temporary bench marks (TBMs).

To ensure accurate primary and temporary bench marks:
- establish primary bench mark* from the agreed Ordnance Bench Mark or from existing works and agree level in writing with the supervising authority
- plan positions of TBMs in good time, taking account of temporary and permanent works (all points of the works should, where possible, be within 40 m of a TBM)
- verify the levels of previously established TBMs by levelling from the primary bench mark
- establish TBMs not more than 80 m apart. Closing error to primary bench mark must not exceed 5 mm
- use existing permanent features for establishing bench marks whenever possible (see illustrations)
- where no permanent feature is available for a bench mark, establish it in firm ground and mark as shown in the illustration
- protect bench marks from site traffic as necessary
- if assumed datum has been used for scheme, as shown on contract drawings, check with the supervising authority that this datum may be used
- record position, reference number, level and date last checked of each TBM and the primary bench mark on the site plan
- display copy of site plan or list of bench marks (with details) in site offices
- check levels of TBMs at regular intervals†
- report any apparent disturbance of TBMs
- update displayed plan/list of TBMs
- transfer levels from TBMs to permanent works as soon as practicable.
- remove redundant TBMs

*BRE levelling station can conveniently be used, *see* Cheney, J. E. in Bibliography *p. 122*
†**Warning:** Earthworks and ground settlement, heave, expansion or contraction can affect TBMs

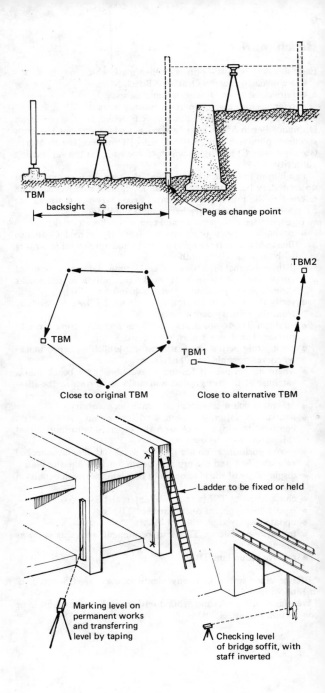

TBM

backsight $\hat{=}$ foresight

Peg as change point

TBM

Close to original TBM

TBM1   TBM2

Close to alternative TBM

Ladder to be fixed or held

Marking level on permanent works and transferring level by taping

Checking level of bridge soffit, with staff inverted

26

Levels

Whether the site engineer is measuring levels (e.g. original ground levels or completed work) or providing levels (e.g. blinding level for slab), the principles are similar.

Using and setting-up an optical level
- check collimation of level at least weekly (two-peg test)   *p. 109*
- ensure firm base and comfortable telescope height
- avoid traffic vibrations
- arrange backsight approximately equal to foresight
- do not leave level unattended!

Use of change points
Use peg, foot plate or other convenient point to:
- avoid sighting more than say, 40 m
- cater for large changes in ground level
- cope with obstructions.

Use of staff
- check that staff is undamaged
- ensure staff is held vertical or 'rocked' (if no bubble is fitted)
- minimise use of extended staff
- if extension used, check catch fully engaged
- avoid sighting on bottom 0.5 m of staff (refraction is severe near ground)
- use inverted staff to measure soffit levels (book reading on staff as negative)
- having marked a given level (e.g. floor level) on a vertical surface (e.g. column face), chainman should remove staff and reset staff to mark before a check observation is made.
- a tape is sometimes substituted for a staff (e.g. in a shaft).   *p. 99*

Use of TBMs
- check for signs of disturbance
- check reduced levels regularly
- transfer to permanent works as soon as practicable
- close series of levels to original or alternative TBM
- check closing error within acceptable limits.

Other levelling methods
- levels can be transferred by spirit level or water level
- a rotating laser can conveniently provide a horizontal   *p. 39*
reference plane but must be regularly checked with an optical   *p. 109*
level.

Use of precise optical level
- for time-dependent deflections e.g. pile load tests
- for precise levelling of mechanical equipment.

For further guidance see Cheney, J. E., Bibliography   *p. 122*

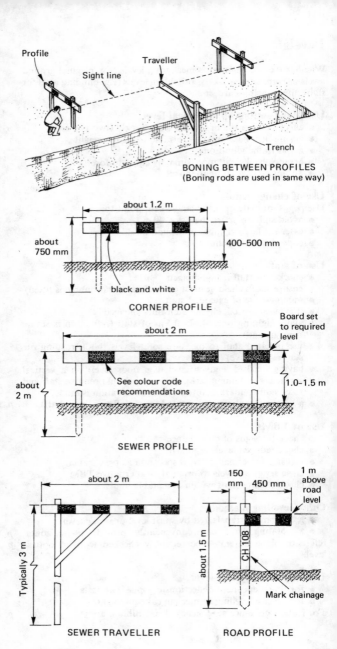

Profile — **Sight line** — **Traveller**

Trench

BONING BETWEEN PROFILES
(Boning rods are used in same way)

about 1.2 m
about 750 mm
400–500 mm
black and white

CORNER PROFILE

about 2 m
Board set to required level
about 2 m
See colour code recommendations
1.0–1.5 m

SEWER PROFILE

about 2 m
Typically 3 m

SEWER TRAVELLER

150 mm
450 mm
1 m above road level
about 1.5 m
CH 108
Mark chainage

ROAD PROFILE

28

Profiles and travellers

Standardisation

It is helpful to standardise as far as practicable on the construction and colour coding of all profiles and travellers. Appropriate sizes are given opposite and a suggested colour scheme is shown on the front and back end-papers.

Use of profiles and travellers

Profiles and travellers are widely used for providing lines and levels. The general principle of 'boning' between profiles or boning rods is illustrated. Specific uses are described later but the general principles are given below.

Sewer profiles and travellers

Sewer profiles are set at a given height above the required trench *p. 45* formation level and to one side of the trench. The horizontal board of the traveller is therefore made to project to that side of the trench so that it can be 'boned through' the profiles.

Each traveller should be marked with its height and the manholes between which it is appropriate (e.g. MH 26–MH 30)

Corner profiles

A pair of profiles is used to define a construction line or parallel *p. 59* lines (e.g. faces of foundation trench and outside face of brickwork). The profile boards may also be set to level and used with a traveller to bone in a required level (e.g. bottom of foundation trench)

The boards should be 'outside' the supporting stakes so that a string line between profiles will not pull the boards off.

Road profiles and travellers

Road profiles are usually provided at intervals along the line of the *p. 81* road, in pairs, one each side of the centre line. The 'chainage' should be marked on each profile (in metres). Initially batter rails are set outside the extent of the earthworks to control the bulk cut and fill operations. When the earthworks are complete new profiles are set just clear of the road itself to control final trimming of the formation and construction of the road. For the latter purposes, a *pp. 85* traveller on a base is helpful, with facility for adjusting the height of *& 87* the board to cater for the various component layers of the road. *p. 81*

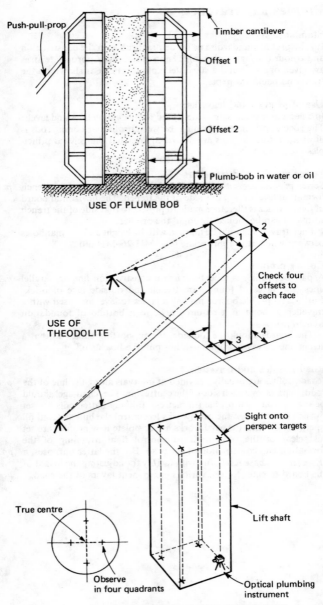

Push-pull-prop

Timber cantilever

Offset 1

Offset 2

Plumb-bob in water or oil

USE OF PLUMB BOB

Check four offsets to each face

USE OF
THEODOLITE

Sight onto perspex targets

True centre

Observe in four quadrants

Lift shaft

Optical plumbing instrument

USE OF OPTICAL PLUMBING INSTRUMENT

Plumbing

Plumbing of modest accuracy can be achieved with a good quality 1-m spirit level, but other methods must be used as the height and/ or need for precision increases.

Use of plumb bob

In the example shown, a freshly concreted wall is checked for verticality. The plumb bob is suspended from a piece of timber nailed to the top of the formwork and shielded from the wind or immersed in a pail of oil or water. Offsets from the back of the form are measured at top and bottom with due allowance for any steps or tapers in the wall. Any necessary adjustments are made with a push-pull prop.

Use of theodolite

The formwork for a tall column form is being plumbed in the example. A theodolite is set up on a plane parallel but offset to one face and sighted on suitable offset marks at the top. (Observe both edges to check on twist.) Similar observations are made on the bottom of the form. Any discrepancy in verticality (mean of observations on left and right face) is read at the bottom for convenience and the column form adjusted. The whole process is then repeated for the adjacent face.

Sighting at a steep angle above the horizontal is facilitated by using a diagonal eyepiece.

Note: The theodolite must be some distance from the column for accuracy; this may be impossible on a cramped site.

Use of optical plumbing instrument

The operation is relatively simple as follows:
- set up and level instrument over ground station
- sight down and centre over ground station
- sight up (through second telescope or by operating prism mechanism) onto target and mark a defined point
- turn instrument through 90°, 180° and 270° in horizontal plane to define three further points
- intersection of diagonals joining four points lies on vertical line through ground point

Optical plumbing is particularly useful for ensuring the accuracy of lift shafts, slipformed structures and climbing forms. The example shows the use of an optical plumbing instrument in a lift shaft, using perspex targets fixed at the top level. At least three ground stations should be used to check for possible twisting.

Use of lasers

Lasers can be used to define a vertical line or plane.

p. 39

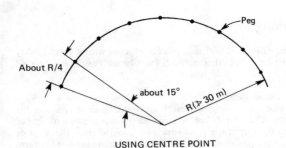

About R/4

about 15°

R(⩾ 30 m)

Peg

USING CENTRE POINT

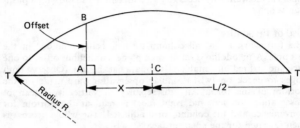

B

Offset

A

C

T

T

X

L/2

Radius R

USING CHORD POINTS

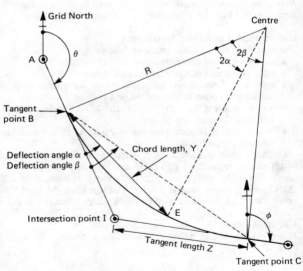

Grid North

A

θ

Centre

R

2α

2β

Tangent point B

Deflection angle α
Deflection angle β

Chord length, Y

E

Intersection point I

Tangent length Z

φ

Tangent point C

USING DEFLECTION ANGLES

32

Circular curves

Using the centre point
Circular curves of 30 m radius and under can be set out by using a steel tape from the centre point, placing pegs equidistant round the circumference of the circle about 15° of arc apart that is, at intervals of about $\frac{1}{4}$ of the radius.

Using chord points
If the centre point cannot be used, a curve can be set out by offsets, given two points on the circle and the radius R:
- measure length of chord, L (if not given)
- set up string line between chord points, T
- set out centre of chord, C
- for distance X along chord from C set out offset AB perpendicular to chord where:

$$AB = \sqrt{(R^2 - X^2)} - \sqrt{(R^2 - (L/2)^2)}$$

Using deflection angles
Large radius horizontal circular curves can be set out with a theodolite, using deflection angles. In the example illustrated, the coordinates of A, B, C, and D and the required radius R are given. In principle, the procedure is as follows:
- from the coordinates of A and B, calculate whole circle bearing (WCB), θ
- for first point on curve (E) choose convenient chord length Y [Suggested chord intervals: Retaining walls 2–5 m. Kerbs 5–10 m. Earthworks 20–30 m]
- calculate deflection angle $\alpha = \sin^{-1}(Y/2R)$
- calculate WCB of E from $B = \theta \pm \alpha$ (minus in this case)
- repeat for further points but note that, for point C, chord length is calculated from coordinates of B and C
- with theodolite on point B set out points on curve.

A worked example is given in Appendix G. *p. 119*

For a long curve or where sights will be obstructed, it will be necessary to set out the curve in sections, moving the theodolite for each section (*see also* Spiral transition curves *and* Appendix H) *p. 77*
p. 120

Note: If intersection point I is given rather than coordinates of C:
- calculate tangent length Z from coordinates of B and I
- calculate deflection angle $\beta = \tan^{-1} Z/R$
- calculate chord length $= 2R \sin \beta$.

Note: Given the grid coordinates of points on a curve, a total station would allow direct setting out. *p. 15*

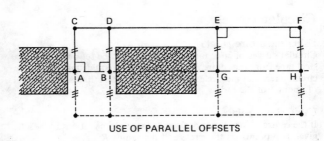

USE OF PARALLEL OFFSETS

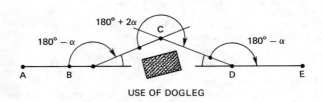

$180° - \alpha$ $180° + 2\alpha$ $180° - \alpha$

USE OF DOGLEG

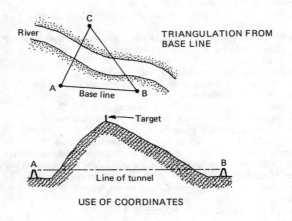

River

TRIANGULATION FROM BASE LINE

Base line

Target

Line of tunnel

USE OF COORDINATES

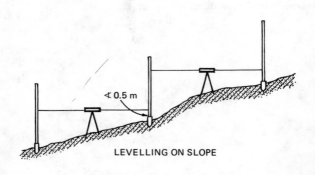

< 0.5 m

LEVELLING ON SLOPE

Obstructions and slopes

All possible problems caused by obstructions and slopes cannot be anticipated but the following hints may help.

Use of parallel offsets
This technique is useful to extend a line on the other side of an obstruction but offset distances must be accurate to minimise potential errors. In the example, C and D are offset from A and B and at right angles to AB. CD is extended to E and F beyond the obstruction. Finally G and H are offset from E and F.

Note: The potential error is magnified because AB is short relative to the length of the obstruction. It would be prudent to set out another offset line on the opposite side of the obstruction and to check the closing error.

Use of dog-leg
This procedure is more elegant but again accuracy is imperative:
- set up theodolite on B, sight on A, traverse $(180° − α)$
- set out C at convenient distance from B
- set up theodolite on C, sight on B, traverse $(180° + 2α)$
- set out D such that CD = BC
- set up on D, sight on C, traverse $(180° − α)$
- set out E to define AB extended.

Note: A total station or theodolite with add-on EDM would further simplify the procedure. *p. 2*

Use of base line
Triangulation from a measured base line can be used to set out points across a river, for example. It may be convenient to have a chainman on each side of the river.

Use of OS coordinates or triangulation
In the example shown, it is most likely that the Ordnance Survey (OS) National Grid would be used to set up stations A and B and to define the bearing of AB. If two points, a reasonable distance apart, can be set up on the hill and seen from both sides, triangulation could be used as an alternative method, with a base line on one side. *p. 111*

Levelling on slope
Back-sights and foresights are necessarily short when levelling up or down a slope. If levelling up the slope, check that the line of collimation is not less than 0.5 m above ground level at the proposed peg location (*see* Use of staff). *p. 27*

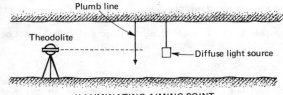

ILLUMINATING AIMING POINT

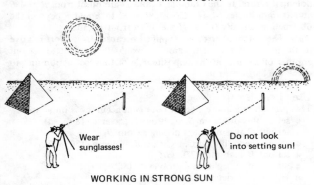

Wear sunglasses!

Do not look into setting sun!

WORKING IN STRONG SUN

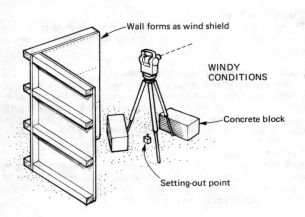

Wall forms as wind shield

WINDY CONDITIONS

Concrete block

Setting-out point

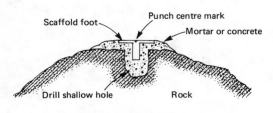

Scaffold foot

Punch centre mark

Mortar or concrete

Drill shallow hole

Rock

SETTING-OUT STATION ON ROCK

Difficult ambient conditions

Poor light
- use surveying instrument with integral lighting system to illuminate scales and cross-hairs
- use diffuse light source to illuminate staff or aiming point (place source behind plumb line, for example).

Strong sun
- shade instrument to avoid bubble disturbance
- wear sunglasses in strong sunlight
- to avoid heat shimmer, set out when sun is low but,
- never look through a telescope into the sun.

Cold weather
- ensure clothing is warm, wind-proof and waterproof
- wear mittens for ease of adjusting instruments.

Windy conditions
- ensure feet of tripod firm or suitably weighted
- in extreme conditions, shield instrument.

Salt spray or dust
- meticulously clean instrument each day (instruct chainman) *p. 11*
- clean lenses with recommended tissue or brush only
- check for bearing wear on instruments frequently. *p. 109*

Noisy conditions
- carry out all preliminary calculations in site office
- brief chainman fully before going into noisy area
- subject to need to hear alarms, wear ear defenders.

Vibrations
- site instruments as far from source as practicable
- set out before operations start or during meal breaks
- ensure readings are consistent before accepting them.

Soft ground
- for setting-out station or bench mark use existing works, *p. 17* construct mass concrete block or install pile
- use offsets liberally; be resigned to resetting pegs *p. 19*
- check peg from offsets, do not assume it is correct
- avoid moving around tripod more than is essential.

Rock
- use rock drill to make hole for level peg or steel pin
- for setting-out station, use scaffold foot as shown.

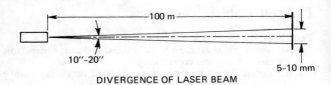

DIVERGENCE OF LASER BEAM

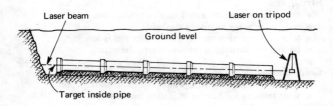

PIPE-LAYING

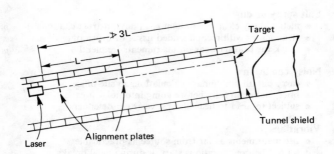

TUNNELLING

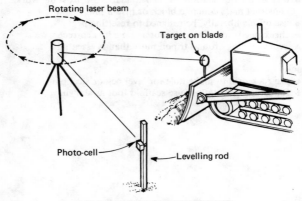

LEVELLING AND EARTHMOVING

38

Lasers

Types of laser

A laser used for setting out may be:
- alignment laser or rotating laser to define plane
- visible-beam (He-Ne) or invisible-beam (e.g. gallium-arsenide)
- Class 1, 2, 3A or 3B as defined by BS 4803 and discussed by Cox (*see* Bibliography) p. 122

Safety and limitations

To avoid eye damage from lasers:
- where practicable, use Class 1 or Class 2 lasers which are effectively low-powered
- if Class 3A or 3B laser must be used, comply with safety procedures recommended by RICS or Cox (*see* Bibliography) p. 122

The minor limitations of lasers are:
- beam divergence—typically 5 to 10 mm per 100 m
- beam is refracted by non-uniform temperatures or where it passes too close to solid material (e.g. tunnel wall)
- invisible beam requires photoelectric detector

In tunnels/ducts ventilate to achieve uniform temperature.

Use of alignment laser

A visible beam is generally most convenient.
For pipe-laying, a laser is set to line and gradient for:
- excavation (using traveller with solid target)
- laying the pipes (using transparent target inside pipe)

For tunnelling, the procedure is usually as follows:
- two plates are fixed to the completed tunnel roof such that small holes in each define a line offset from the tunnel centre (set by standard setting-out techniques)
- a laser is fixed to the tunnel roof and adjusted until the beam passes through the holes in the two plates
- the beam projects onto a target on the tunnel shield or tunnelling machine for control of line and gradient
- the laser is moved forward and reset when the target distance reaches maximum for accuracy (see figure).

Use of rotating laser

Both visible-beam and invisible-beam lasers are available, but the latter may be favoured to avoid the potential nuisance of a visible occulting beam.

All rotating lasers should be gyro-compensated and able to operate in the horizontal or vertical plane. Typical uses include:
- earth-moving to level on horizontal plane
- providing a horizontal reference plane across a site
- providing a vertical reference plane for curtain walling.

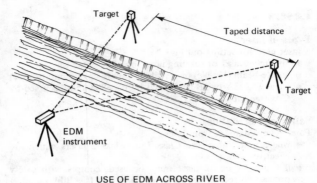

USE OF EDM ACROSS RIVER

USE OF EDM OVER HILL

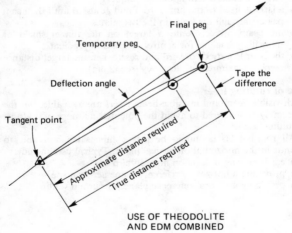

USE OF THEODOLITE
AND EDM COMBINED

Electromagnetic distance-measurement

Basic principles

Electromagnetic distance-measurement (EDM) is based on measuring the transit time of an electromagnetic beam from a transmitter to a reflecting prism and back again. The direct distance is displayed digitally but may need small corrections for ambient conditions. A slope correction must be made if the transmitter and reflector are at different heights. The equipment can be used separately for distance measurement alone or mounted on a theodolite so that horizontal angles can also be measured. The claimed accuracy is $\pm (5\,mm + 5\,mm/km)$ or better.

Note: On some models, a correction factor for ambient conditions can be input and slope is electronically detected. Both ambient and slope corrections are then automatically applied.

Possible applications

Apart from setting out critical lengths (e.g. long base lines), other applications include:

- measurement of length where taping is impracticable, e.g. over water
- measurement of length over hilly terrain (with due correction for slope)
- setting out or checking coordinates (when combined with theodolite), including coordinates of curves.

Setting out lengths

To allow for slope correction, proceed as follows:

- set out temporary peg with/without slope correction,
- measure distance accurately with slope correction, making any necessary corrections for ambient conditions and instrument and reflector geometry
- subtract the corrected value from the required length
- tape the difference from the temporary to final peg
- check distance to final peg, with slope correction.

Note: Some reflectors display measured data transmitted from the distance meter.

Checking EDM instrument

Check EDM instrument at least weekly against known length (e.g. a base line) set out over reasonably level ground by taping with all corrections applied. *See also* Appendix B.

Warning: Nickel–cadmium batteries should not be recharged until p. 109 nearly, or preferably fully discharged so keep check of cumulative use. Discharge any residual charge according to manufacturer's instructions before recharging overnight.

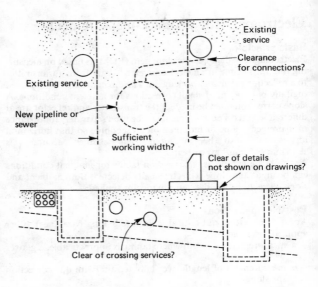

Existing service

Clearance for connections?

Existing service

New pipeline or sewer

Sufficient working width?

Clear of details not shown on drawings?

Clear of crossing services?

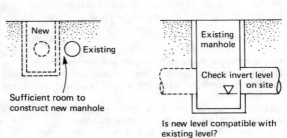

New

Existing

Sufficient room to construct new manhole

Existing manhole

Check invert level on site

Is new level compatible with existing level?

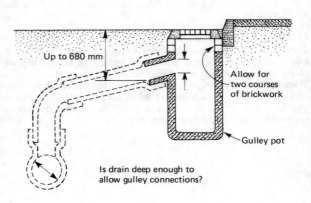

Up to 680 mm

Allow for two courses of brickwork

Gulley pot

Is drain deep enough to allow gulley connections?

Sewers and drains: initial checks

The site engineer should mark the approximate line of the sewer and then:

- walk along the planned line, noting existing features
- note any potential setting-out or construction problems
- where possible, check likely soil conditions
- clear any discrepancies with the supervising authority before proceeding with the detailed setting-out.

Adjacent and crossing existing services

The supervising authority should have notified the other statutory undertakers of the intended sewer works, obtained details of adjacent services and checked under the *Model Agreement* (*see* Bibliography) that adjacent gas mains will not be at risk.

p. 122

Consult public utilities' drawings of mains and cables to ensure that proposed sewers and manholes should not conflict with these services. If in doubt, dig trial holes.

Do not assume that an existing sewer follows a uniform gradient or straight line between manholes. Do not forget to take into account pipe thickness and collars or patent joints.

Mark up drawings with all available information and confirm these with statutory undertakers.

Check that the distance between new and existing services will be sufficient for excavation, trench support and working space, including construction of manholes.

Warning: Record drawings of existing services are frequently inaccurate.

Interference by drain connections

Inspect the drawings for the location of connections, and determine whether these will interfere with any adjacent service runs, particularly as regards level.

Note: House services are unlikely to be shown.

Manhole positions

Agree and confirm with the supervising authority the intended location of the manholes on the ground. Normally, sewer drawings do not show exact positions or take account of local details.

Discharge Levels

If a new sewer is to discharge into an existing manhole or outfall, check physically on site that the level shown on the drawings is accurate. This is vital; errors may require redesign of part or whole of the system.

Gully connections

Check that the surface water sewer invert is such that connections from road gullies can be made. The distance below the top of the cast iron gully grate to the invert of the outlet pipe from the gully pot will be up to 680 mm.

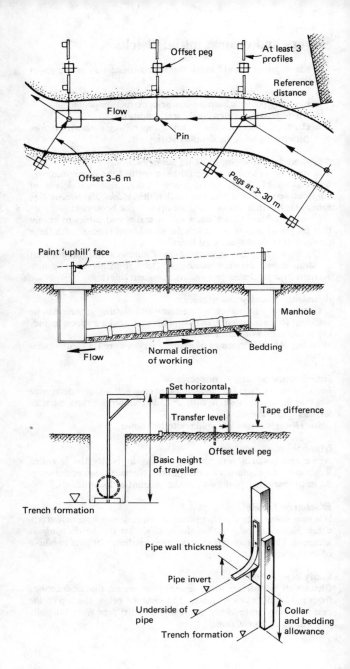

Offset peg

At least 3 profiles

Reference distance

Flow

Pin

Offset 3–6 m

Pegs at ⊁ 30 m

Paint 'uphill' face

Manhole

Flow

Normal direction of working

Bedding

Set horizontal

Transfer level

Tape difference

Offset level peg

Basic height of traveller

Trench formation

Pipe wall thickness

Pipe invert

Underside of pipe

Trench formation

Collar and bedding allowance

Sewers and drains: line and level

Centre-line
Set out the centre-line of the trench with pegs or road pins not more than 30 m apart. Mark the centre of each manhole with a peg, referenced to other pegs or nearby permanent features. *p. 19*

Offset pegs
Position offset pegs or road pins 3–6 m on one side of the centre line. The offset distance should take account of where excavated material will be placed and the space required for movement of plant but should be constant between any pair of manholes. *p. 19*

Profiles
Set up standard profiles on the same side of the centre-line as the offset pegs. If possible the end of the profile board should be not more than 3 m from the centre line. Paint 'uphill' face of profile board (see illustration; for colour code see end-papers). *p. 29*

Find level of offset peg and calculate depth to the pipe invert or the trench formation level before deciding the height of the traveller (see Traveller, below). Calculate the distance from the offset peg to the top of the profile board and fix board by transferring peg level to stake and measuring upwards as illustrated. Indicate on profile its height above invert level or trench formation as appropriate.

Avoid use of traveller less than 2 m high. Provide one profile at each manhole and at least one between manholes. The middle profile(s) allows a quick check on the accuracy of the other two. Calculate the level(s) for the middle profile(s) from the levels and gradient shown on the drawings; do not set up by boning through the other profiles.

Traveller
The basic height of the traveller should be the difference in level between the line defined by the profiles and the pipe invert or the trench formation level as appropriate, and should be a convenient multiple of 0.5 m. Excavation level, bedding level and invert level can all be indicated, if wished, as in the sketch. Mark traveller with its height and manholes between which it is to be used. *p. 29*

Information sheet
Provide foremen/gangers with site information sheets recording relevant information on pegs, profiles, travellers and bedding. *p. 53*

Suggested colour code
A suggested colour code for pegs and profiles is shown on the end-papers.

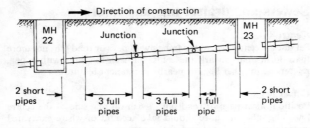

LOCATION OF JUNCTIONS

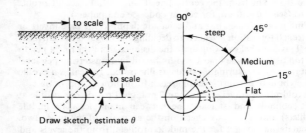

ANGLE OF JUNCTION IN SECTION

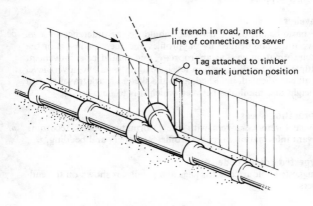

MARKING BEFORE BACKFILLING

Sewers and drains: junctions

Establishing location

- Show the plan location of each junction on a sewer information sheet (and site information sheet as appropriate). *p. 53*
- Express the location of each junction in terms of the numbers of pipes to be placed from a given manhole before the junction, or the number between junctions.
- Tape position for later fitting of saddle connections

Note: Specifications frequently require that the pipe built into the manhole and the next pipe be short pipes to avoid excessive bending load on the pipe. In the example shown, the first junction is placed after 5 pipes (2 short and 3 full) and there are 3 full pipes to the second junction.

Estimating angle

- Estimate angle of junction in plan and specify type of junction required on information sheet (catalogue number can be used)
- Estimate the angle of each junction in section and classify as flat, medium or steep as illustrated.
- Provide foreman and ganger with site information sheet showing definitions of flat, medium and steep.

Note: If angles are difficult to determine, seek the supervising authority's agreement to use saddle connections.

Marking before backfilling

Mark junctions before backfilling for ease of constructing trenches to make connections.

- Instruct the foreman or ganger to mark each junction before backfilling as illustrated
- Measure the 'as laid' location of each junction by taping from the inside face of the nearest downstream manhole
- Record on 'master copy' of sewer information sheet.

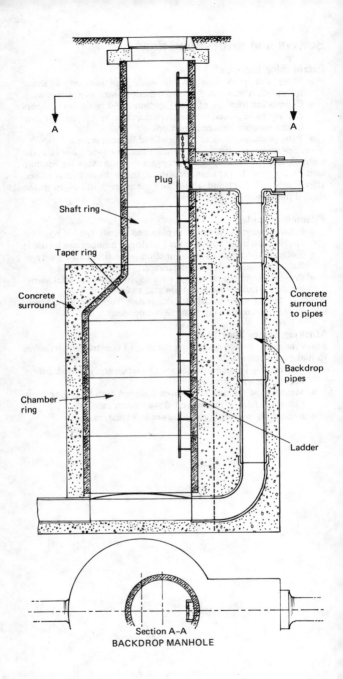

Plug

Shaft ring

Taper ring

Concrete
surround

Chamber
ring

Concrete
surround to pipes

Backdrop
pipes

Ladder

Section A–A
BACKDROP MANHOLE

Sewers and drains: manholes

Manholes in general

A gang separate from the pipelaying gang(s) may be used to construct the manholes. Information relating to a manhole is therefore best provided to the foreman/ganger on a separate site information sheet.

p. 53

The following should be included as appropriate:

- base location, dimensions and level
- type and number of concrete rings (if used)
- taper
- invert level
- channel and benching details
- concrete surround (if required)
- cover slab level
- location of man-entry hole and step-irons or ladder
- any divergences from typical manhole details.

If concrete rings are used:

- designate each specific type by a code letter (to correspond with information sheet)
- paint code letters on rings when delivered.

This helps to ensure correct order of construction and overall depth

Back drop manholes

When dealing with back drops in a stretch of sewer, errors often arise due to the abrupt changes in depth. To avoid errors, adopt the following procedure:

- Draw attention to manhole on the sewer information sheet
- Set up the profile posts for the section containing the higher run of sewer, but do not fix the profile board until the preceding downstream section is excavated.
- When the higher section is ready for excavation, withdraw the traveller, re-set the profile and issue the new traveller. Details must be shown on a site information sheet.
- Provide details of the construction of the back drop and manhole to the foreman/ganger on a site information sheet.

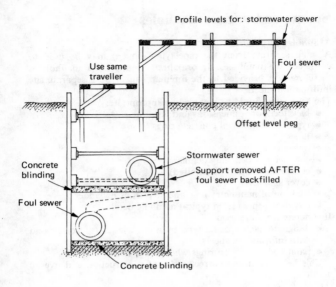

Profile levels for: stormwater sewer

Use same traveller

Foul sewer

Offset level peg

Concrete blinding

Stormwater sewer

Support removed AFTER foul sewer backfilled

Foul sewer

Concrete blinding

COMMON TRENCH

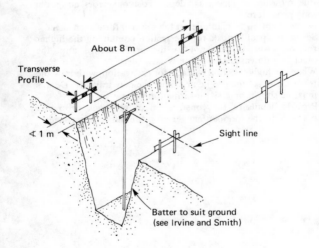

About 8 m

Transverse Profile

< 1 m

Sight line

Batter to suit ground (see Irvine and Smith)

DEEP TRENCH (especially with battered sides)

Sewers and drains: common and deep trenches

Common trench
It may be convenient and economic to lay both foul and storm water sewers in a common trench but at different levels. Note, however that the support must be appropriate for the greatest depth (consult *Irvine and Smith, see* Bibliography). The walings and struts must be arranged to suit the construction of both sewers conveniently and with safety. On a sketch to the foreman/ganger, show clearly any temporary strutting and when it may safely be removed. p. 122

Before setting out check that the invert level of upper sewer will allow connections to the lower sewer. To set out the sewers:

- issue a site information sheet to the foreman/ganger
- erect upper and lower profiles as shown and paint 'uphill' face of the sight rails (see end-papers for colour code). p. 44
- ensure that trench sheeting will not obstruct boning on lower sight rail
- construct the traveller with one cross piece which will be used in conjunction with the upper or lower sight rail as appropriate.

Deep trench
In a deep sewer trench with battered sides, the usual profiles cannot be set within 3 m of the centre-line, and accurate boning is not possible. Proceed as follows:

1. Set up pairs of profiles *parallel* to and on both sides of trench at about 8 m intervals as illustrated.
2. Bone-in traveller across trench to give appropriate excavation level, leaving ground a little high.
3. Set up new profiles perpendicular to trench line and within trench for final trimming to level.
4. Set up temporary datum level within trench to set pipes to level.

In a deep supported trench, it is difficult to obtain accurate levels when the traveller is more than 5 m high, therefore:

1. Excavate roughly to depth using normal profiles.
2. Set up profiles in trench and trim to final level.

Note: Beware disturbance of profiles in trench if trench sheets or sheet piles are being driven progressively. Check after each round of driving.

SEWER INFORMATION SHEET

Date: Copies to:
Contract:
Compiled by:

PIPE:	UPSTREAM MANHOLE:
Class	Manhole No.
Type	Type:
	Standard/Backdrop/Catchpit
Diameter mm	Construction
Gradient	External dimensions mm
TRAVELLER:	Internal dimensions mm
Height (A) mm	Surround
	Backdrop/Catchpit level

UPSTREAM
Invert level (B)
Offset peg level (C)
Profile board height
above peg ($A + B - C$) m
Ground level (GL)
GL to invert   m

INTERMEDIATE
Invert level (D)
Offset peg level (E)
Profile board height
above peg ($A + D - E$) m
Ground level (GL)
GL to invert   m

DOWNSTREAM
Invert level (F)
Offset peg level (G)
Profile board height
above peg ($A + F - G$) m
Ground level (GL)
GL to invert   m

Peg No.   Peg No.
Peg No.   Peg No.
Peg No.   Peg No.   Distances from D/S MH

PIPE BEDDING:	DOWNSTREAM MANHOLE
	Manhole No.
Material	Type:
Thickness mm	Standard/Backdrop/Catchpit
	Construction
Width mm	External dimensions mm
	Internal dimensions mm
	Backdrop/Catchpit level

Notes:
1. Insert distance from centre-line to offset peg, L or R.
2. Insert existing services (in red) and distances from D/S MH.
3. Insert potential obstructions and distances from D/S MH.
4. Indicate junction by short line to L/R, add diameter, angle of junction (flat/medium/steep), and number of junction in string from D/S MH.

Sewers and drains: information sheets

Sewer information sheet

The purpose of this sheet is to ensure that site engineers and general foremen concerned with the supervision of the work are fully acquainted with what is required. The sheet also forms a record for future valuation purposes. To assist the site engineer, the sheet is designed to be comprehensive for the majority of situations. As many 'boxes' as possible are provided for data to minimise omissions. Note that the direction of flow is fixed (top to bottom on the sheet). The manhole numbers must be inserted accordingly and also the direction of working; this is usually in the direction opposite to the flow. (It is difficult to lay pipes downhill, especially those with collars)

Given that all the boxes or options are completed, the site engineer need only add the following:

- distances of offset pegs (alternative pegs provided on left and right of centre-line on sketch)
- known existing crossing services (mark in red)
- distance from downstream manhole to existing services or other potential obstructions across the line of the trench (to nearest 0.5 or 1 m should be sufficient)
- approximate lines of existing services parallel or nearly parallel and close to the line of the new sewer
- for junctions: diameter, location (number of pipes from downstream MH), direction (draw short line to left or right of centre-line) and angle (flat, medium or steep)   *p. 47*

In the example opposite, reduced from an A4 sheet, insertions are handwritten.

Site information sheet

The purpose of a site information sheet is to provide the foreman/ganger with just sufficient information for the immediate job in hand. The information can be extracted from the relevant sewer information sheet and any necessary notes added, for example, relating to trench support.

Sketches should be drawn on a site information sheet so that a record will be retained in the site office.

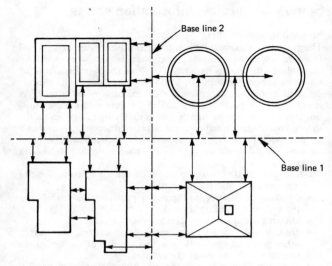

DIAGRAM OF SEWAGE DISPOSAL WORKS

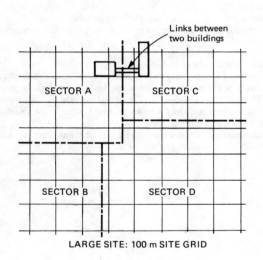

LARGE SITE: 100 m SITE GRID

Buildings: location

Buildings may also be taken to apply to a number of civil engineering structures; the techniques and principles are broadly similar.

Buildings on same site

A project may involve several buildings or structures on the same site. The site engineer must ensure that individual units relate to each other, so that the whole site is laid out in accordance with the drawings. This is particularly important when pipeworks or services connect one with another. It is vital that the structures are set out to the tolerances specified for each structure.

For a small site a simple system of base lines and offsets can be used as shown in the diagram opposite of a sewage disposal works. *p. 13*
Two base lines at right angles are sufficient in this case.

Large site

A very large site with a number of related structures poses additional problems which are mainly organisational. The site engineer in this case may find himself one of a team, each engineer being responsible for the setting out and control of a structure or part of a structure. Under these circumstances, it is essential for one engineer to co-ordinate the overall setting-out activity.

On a large site, there is no substitute for a grid of the whole site with a series of reference stations established round the site so that every part of a structure can be defined by coordinates and bearings to the stations. *p. 15*

Each site engineer must:
- have a layout drawing of the whole site, showing the related structures, grid lines and main setting-out stations
- prove the main setting-out stations he will be using *p. 115*
- check his setting out by reference to structures or points already set out on adjacent sectors of the site
- liaise closely with the site engineers for adjacent sections where structures must be set out by direct reference to one another, both for location and level
- report immediately any apparent discrepancies in the site grid or between adjacent sections.

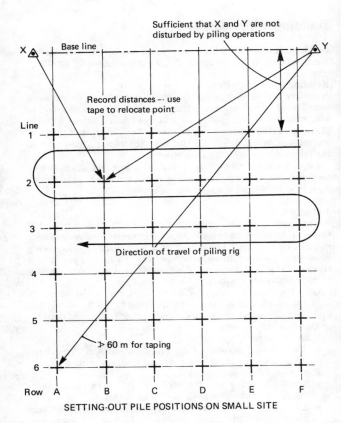

SETTING-OUT PILE POSITIONS ON SMALL SITE

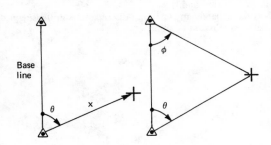

USE OF BEARING AND DISTANCE OR
INTERSECTION ON LARGE SITE

56

Buildings: piling

Responsibilities

Piling may be sublet to a specialist piling subcontractor. The subcontractor may then ask the main contractor to do the setting out or may prefer to use his own engineer. In either case, the final responsibility for correctly setting out remains with the main contractor.

Major problems in setting out

- driving displacement piles causes soil disturbance
- vibrations from piling can disturb instruments
- spoil from boring may be in the way
- piling rigs need space to manoeuvre
- most piles have to be set out individually
- the site engineer must be constantly available

Setting out on small/medium site

Where the number of piles is modest:

- set steel pins to the required positions, using a bricklayer's line and steel tape, from the corner profiles or site grid
- define a suitable base line and record distances to each pin from the two stations
- remove steel pins
- re-locate each pile position when piling rig is ready

Setting out on large site

When the site is large and the pile positions spread widely, it may be more appropriate to locate pile positions by bearing and distance or by intersection from a base line. Check critical pile positions, e.g. at the ends of a row of piles, by redundant observation. For example, use two angles and one distance.

Levels

For cast-in-place concrete piles:

- install temporary level peg by completed bore
- check founding level by using weighted steel tape
- give ganger dimension from peg level or top of casing to concreting level (**not** cut-off level).

After installation

Before piling rig is removed from site:

- check construction tolerances of piles
- report any discrepancies on the pile log sheet
- agree remedial action with the supervising authority.

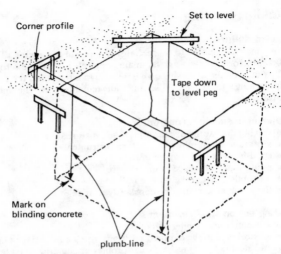

NOTE: Support to excavation omitted for clarity

TRANSFERRING LINES AND LEVELS TO BOTTOM OF EXCAVATION

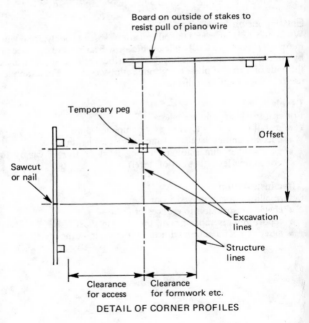

DETAIL OF CORNER PROFILES

Building: excavations

If topsoil is first to be removed, set out in two stages.

Stage 1: Removal of topsoil
- determine overall excavation dimensions
- set out temporary pegs defining area to be excavated
- sprinkle sand along string line between pegs as guide.

Stage 2: Main excavation
- set up stakes for corner profiles offset from structure lines by a convenient multiple of 1 m.
- set and level peg adjacent to each profile
- set profile boards, to a level such that the height of traveller for excavation is multiple of 0.25 m (traveller preferably not more than 5 m high)
- check that clearance between profiles and excavation lines is sufficient for plant access
- set out structure lines by marking opposing profiles
- set out excavation lines from corner lines by distance needed to give clearance for formwork etc.
- provide foreman/ganger with dimensioned sketch on a site information sheet showing profiles, excavation lines, structure lines and traveller height.

Structure lines
Shallow excavations:
- use theodolite if sufficient formation is visible

Deep excavations:
- set up piano wire between opposing profiles
- plumb down line as close to corners as practicable.

Levels
Excavations up to 5 m deep:
- bone between profiles to give four edge strips to level
- use boning rods to level dumpling
- set pegs for blinding by levelling from TBM

Excavations greater than 5 m deep:
- set up board across corner of excavation to a reduced level
- tape down to level peg at bottom of excavation
- set up temporary profiles within excavation
- set pegs for blinding by retaping
- agree peg levels with the supervising authority or seek other independent check.

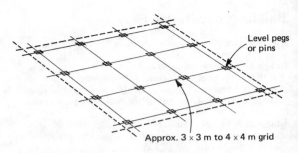

Level pegs
or pins

Approx. 3 × 3 m to 4 × 4 m grid

PEGS FOR BLINDING CONCRETE

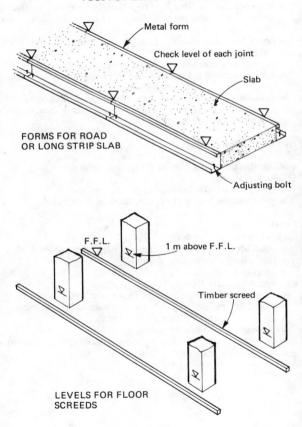

Metal form

Check level of each joint

Slab

FORMS FOR ROAD
OR LONG STRIP SLAB

Adjusting bolt

F.F.L.

1 m above F.F.L.

Timber screed

LEVELS FOR FLOOR
SCREEDS

Buildings: slabs and floors

Ground slabs

- set up profiles to control excavation to formation level
- drive 300 mm steel pins to level on a grid of between 3 and 4 m centres to control blinding.

If ground slab is to be screeded, the edge form controls the structural thickness of the slab. Minor variations in level are absorbed within the screed thickness.

If no screed is incorporated, set timber or metal forms to level to control finished floor level (FFL). Metal forms are preferable for accurate results and frequently incorporate adjusting bolts. Check level at each joint between forms.

Precise levelling may be necessary for warehouses where the use *p. 27* of fork lift trucks demands high accuracy.

Suspended slabs

Where there are columns at regular intervals, mark convenient reduced level near tops of columns. Carpenters can then level timber bearers adjacent to columns and determine intermediate bearers with string line.

If there are no intermediate columns (e.g. diagrid slab), set levels of selected timber bearers allowing for thickness of sheathing. Similarly, if soffit is curved.

If the suspended slab is massive or spans are unusually large, check whether the specification requires any precamber.

An allowance should be made for deflection of the falsework under the concrete load, where the slab is massive or the supporting falsework is greater than storey height.

Carry out spot checks when formwork is complete, preferably before reinforcement is fixed. Reinforcement levels may need checking on deep sections to ensure cover is correct.

Screeds

Mark columns 1 m above FFL. Screed layers will set timber screeds from these levels. Carry out spot checks on timber screeds before screed is laid.

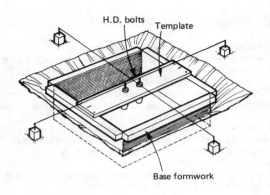

STANCHION BASE

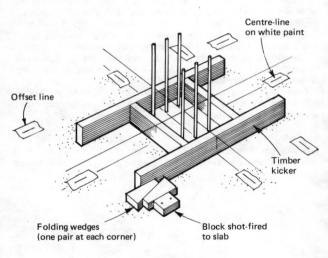

COLUMN KICKER

Buildings: bases, columns and walls

Stanchion or column bases

- set out the centre-lines of the stanchion or column with four pegs outside the area excavated for the base
- when the base formwork has been fixed in relation to these pegs, set up theodolite afresh, check pegs and mark centre-lines on plywood or timber template nailed across top of form. Holding down (HD) bolts, mortice boxes or column kickers are set out from centre-lines by carpenters
- level from TBM to mark concreting level on form
- check that formwork or template is not apparently disturbed during placing of concrete (use pegs)
- carry out final check on template position after concrete is placed but before final set (use theodolite).

Column and wall kickers

The centre-lines of columns or walls and convenient offset lines should be set out on a previously cast slab so that the carpenters can locate the kicker form. It is imperative to set out these lines as soon as the slab has hardened and before it is used for stacking reinforcement, formwork and other materials. Ensure that the marked lines will remain accessible.

Note the illustrated use of timber blocks shot-fired to the slab to hold the form in position with folding wedges.

Column and wall formwork

- check that the kickers are aligned with the marked centre-lines and are square in both plan and elevation
- check formwork for verticality with plumb line or theodolite immediately before placing concrete *p. 31*
- provide concreting level from TBM, taping vertically as necessary
- check verticality of formwork after concrete is placed but before final set.

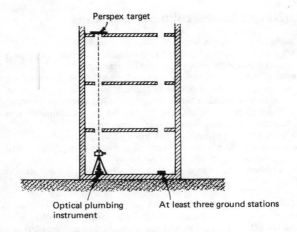

PLUMBING MULTI-STOREY BUILDING

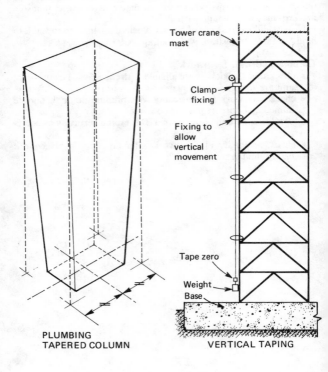

PLUMBING TAPERED COLUMN

VERTICAL TAPING

Buildings: tall structures

Plumbing

The primary instruments for plumbing are:
- plumb bob
- theodolite
- optical plumbing instrument.

The principles of use are described under Plumbing. *p. 31*

Verticality and twist

Tall structures must be checked for verticality and twist. This is best achieved by plumbing up or down from four points.

Rectangular structures

The use of an optical plumbing instrument to control the verticality of a multi-storey building is illustrated.

Within multi-storey buildings, the plumbing of lift shafts is a particularly critical operation because the installation tolerances are small. Four setting-out points should be established at the base of the lift shaft such that the vertical lines through them will not be obstructed by formwork or scaffolding. Plumbing can be from top to base using plumb bobs or from base to top using an optical plumbing instrument. The latter is preferable if a plumb bob would be disturbed by winds. *See* illustration under Plumbing. *p. 31*

Tapered structures

To plumb a tapered column or similar structure:
- set out orthogonal centre-lines on the base
- plumb from top corners (plumb bob or theodolite)
- check equality of offsets on all four sides

If the structure narrows toward the top, it will be necessary to cantilever out from the top of the formwork to fix plumb bobs.

Height and level

Floor-to-floor dimensions are controlled by a weighted steel tape, measuring each time from a datum at the base of the structure. Each floor is then provided with datum marks in key positions from which to transfer levels on each floor.

The base datum level should be set in a location which allows unrestricted taping to roof level. If a tower crane is used, a tape can conveniently be fixed to the mast (*see* illustration)

Warning: Errors, apparent or real, can result from differences in thermal movement of the tape relative to that of the building, particularly if construction spans a number of seasons.

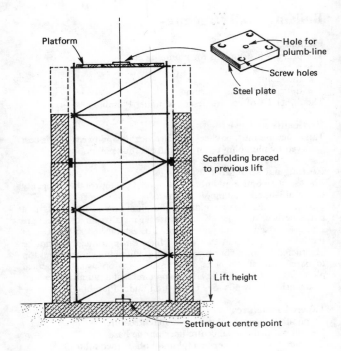

Platform

Hole for plumb-line

Screw holes

Steel plate

Scaffolding braced to previous lift

Lift height

Setting-out centre point

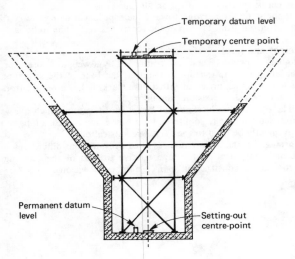

Temporary datum level

Temporary centre point

Permanent datum level

Setting-out centre-point

Building: circular structures

Constant radius

The simplest circular structure is where one or both of the internal and external diameters are constant. Setting out such a concrete structure is relatively simple:

- establish a setting-out centre point on the base
- set out kicker by taping and place kicker
- erect and plumb formwork and place first lift *p. 31*
- strip formwork
- erect scaffold tower braced against first lift of concrete with top platform level with top of second lift
- establish temporary centre point on platform by plumbing up or down from setting-out point
- set up second lift of formwork
- check formwork radius from temporary centre point
- repeat sequence for further lifts.

Varying radius

The same basic procedure as above is followed except that it is necessary to calculate the radius at a given height and to control that height. The sequence of operations is amended as follows:

- erect scaffold (as before)
- establish temporary setting-out point (as before)
- establish temporary datum level on top scaffold platform by taping from permanent datum level
- set up next lift of formwork to level
- check radius of top of formwork.

Establishing temporary centre points

If a plumb bob is used, suspend the bob from a line or piano wire passed through a steel or plywood plate. The plate can be screwed to the top platform when the plumb bob is centred over the setting-out point.

If an optical plumbing instrument is preferred, sight up onto a perspex target fixed to the top platform. *p. 31*

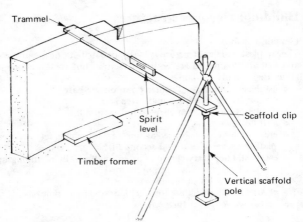

CIRCULAR BRICK WALL

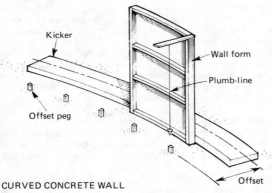

CURVED CONCRETE WALL

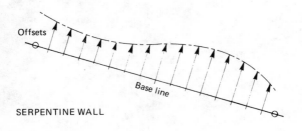

SERPENTINE WALL

Buildings: curved walls

Circular wall—centre accessible
If the centre is accessible and the radius not more than say, 30 m, it is convenient to set out the first course of a brick wall by taping from the centre (similarly for the kicker of a concrete wall). The bricklayer continues by plumbing vertically with a spirit level and checking segments of the curve with a timber shaped to the radius.

If the centre is available and the radius is less than say, 5 m, a scaffold pole can be set up as the centre for a trammel board as illustrated. The inner and/or outer radius are marked on the trammel which is raised course by course with a scaffold clip to control the level of the wall.

Circular wall—centre not accessible
Where the centre is not accessible, the curves must be set out using deflection angles and chord lengths. Chord intervals should be about 3 m. *p. 33*

Taking a concrete wall, for example, the procedure is:
- set out centre-line
- establish offset pegs
- erect profiles, excavate and place foundation
- set out kicker accurately from offset pegs
- place kicker
- set up forms and plumb down to offset pegs or secondary offset points.

Serpentine walls
Serpentine or S-shaped walls are difficult to set out accurately by using deflection angles and chord lengths. An alternative worth considering is to use a base line and offsets, provided the offsets lie within a convenient range for taping. Offset pegs can conveniently be established on the base line at suitable intervals. The offset distances can be calculated provided the geometry of the wall is known.

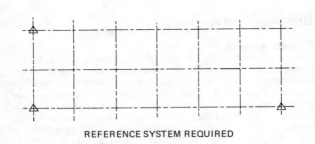

REFERENCE SYSTEM REQUIRED

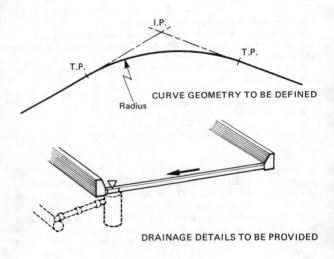

CURVE GEOMETRY TO BE DEFINED

DRAINAGE DETAILS TO BE PROVIDED

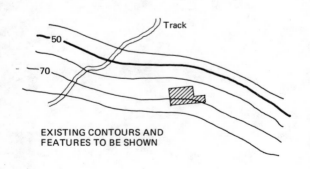

EXISTING CONTOURS AND
FEATURES TO BE SHOWN

Roads: checking drawings

Initial actions
The initial actions described previously should be carried out. They p. 9
include a review of the drawings together with some of the general
points to be checked. Checks on information specific to roads are
given below.

Information to be checked
It is more likely that information may be less detailed for estate and
similar minor roads than for major roads. Nevertheless apply the
checks to all drawings as appropriate:

- where OS triangulation points and bench marks are to be used,
  coordinates and levels should be available on current OS
  listing and must agree with drawings p. 111
- reference points, a base line, or a site grid should be provided.
  If not, a suitable reference system must be agreed with the
  supervising authority
- details of existing features should be provided. Check these
  against latest OS large-scale map and by walking the site
- existing contours or a grid of existing ground levels should be
  shown
- check that fence lines are accurate
- check that overall width between fence lines is sufficient to
  construct the earthworks (*see also* Dealing with discrepancies) p. 73
- borrow pits and tipping areas should be defined or agreed.
  Check that these are adequate for the volume of earth to be
  removed or dumped
- access points to site, borrow pits and tipping areas should be
  defined
- horizontal curves should be fully defined with tangent points,
  etc. If not, design and agree suitable curve with supervising
  authority
- gradients, vertical curves, crossfall or camber and levels at
  junctions should be defined or agreed.
- road drainage details should be shown. Check that there are
  adequate falls to gullies and along drains. For connections into
  existing drains, inverts should be stated and these checked on
  site.

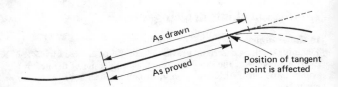

DISCREPANCY IN LENGTH

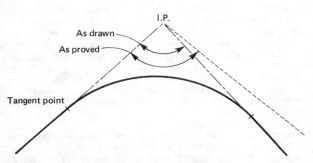

DISCREPANCY IN INTERSECTION ANGLE

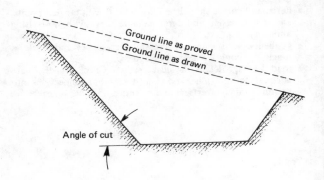

NOTE: Volume and width of cut are affected
DISCREPANCY IN EXISTING GROUND LEVELS

Roads: general procedures

Proving the survey

For major road schemes, the survey and establishment of the road centre-line may be carried out by specialist surveyors. In this case it should only be necessary to ensure that all the primary lines and levels have been established and that existing features agree with those indicated on the drawings.

For other road works, it is essential to 'prove' the survey against the drawings before starting any setting-out:

- prove main setting-out stations *p. 115*
- check or establish and agree primary and temporary bench marks *p. 25*
- establish intersection points on curves
- measure intersection angles
- measure tangent straights and chainages
- check position and coordinates of intersection and tangent points by reference to main setting-out stations
- check chainages at natural features, e.g. hedges
- check relative position of other existing features
- check existing ground levels along centre-line at suitable intervals (use a suitable grid for wide roads or on steep side slopes)
- check whether any existing feature or temporary works may obstruct setting out.

Dealing with discrepancies

Discrepancies typically affect:

- road lengths
- intersection angles
- existing ground levels

To minimise delay:

- double check that there is a discrepancy
- report discrepancy and put forward proposed solution
- agree future action with supervising authority.

Note: Corrections to take account of discrepancies frequently require significant changes to landtake, gradients, crossfalls and drainage details.

Checking computer printout

If a computer printout for setting out the road centre line is provided, check that the changes in coordinates appear logical and that the ends of the section of road tie in with known features. See Appendix H. *p. 120*

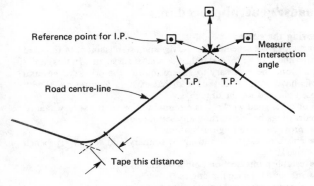

ESTABLISHING MAIN POINTS ON CENTRE-LINE

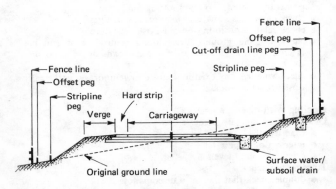

ESTABLISHING LINES BEFORE EARTHWORKS
(Typical rural road)

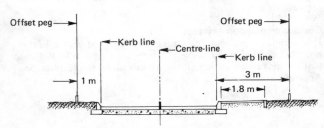

ESTABLISHING KERB LINES AFTER EARTHWORKS
(Minor access road)

Roads: lines

Establishing the centre-line

Set out the centre-line generally in the following order:

- establish intersection points   *p. 19*
- provide offset reference pegs to intersection points
- measure intersection angles
- establish tangent points of curves by taping from intersection points
- set out curves.

Other lines and offsets

From the centre-line set out:

- fence lines
- drains and ditches
- site stripping area
- main chainage offset pegs (beyond limits of earthworks but within fence lines).

Use standard offsets where possible. Record all relevant dimensions on site information sheet.   *p. 7*

After earthworks complete

- re-establish centre-line from offsets
- check intersection and tangent points with respect to main setting-out points
- set out kerb lines
- set out offset pegs convenient distance beyond kerb line or footway.

Widening existing roads

Use the road surface for setting out the centre-line where appropriate. Road nails or shot-fired nails should be used to define the permanent line from which to set out offset pegs.

Pins should be numbered in white paint on the road surface and the following shown on an information sheet:

- pin number
- chainage
- offset distance to point 1 m beyond face of new kerb.

Provide separate standard level pegs or pins.

Suggested colour code for pegs

A suggested colour code for pegs (and profiles) is shown on the endpapers.

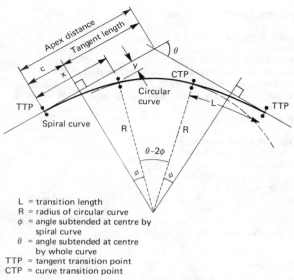

L = transition length
R = radius of circular curve
ϕ = angle subtended at centre by spiral curve
θ = angle subtended at centre by whole curve
TTP = tangent transition point
CTP = curve transition point

BASICS OF SPIRAL TRANSITION CURVE

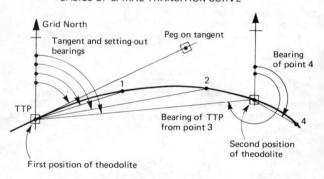

SETTING OUT CURVE

Roads: horizontal curves

Circular curves
Many minor roads incorporate simple circular curves only. These
may be set as described previously, usually by using deflection
angles. *p. 33*

Major roads designed for high-speed traffic, however, require
spiral transition curves, see below.

Specification of spiral transition curves
It is usual for the supervising authority to specify the radius and the
appropriate transition curve table to be used. The contractor may
also be supplied with the deflection angles and chainages in the
form of a computer print-out.

If this is not the case, setting out spiral transition curves is now
greatly simplified by the Highway Transition Curve Tables compiled by the County Surveyors Society (*see* Bibliography), given the *p. 122*
radius and the appropriate Table No. by the supervising authority.

Setting out transition curves
The procedure is similar to that for setting out horizontal circular
curves by using deflection angles. It will be necessary to reposition *p. 33*
the theodolite for long curves. Given the table of bearings and
chainages, provided or calculated, the procedure is as follows:

- set up the theodolite on the origin of spiral (TTP)
- set the known bearing of a setting-out station (or other suitable
  point), align theodolite on that point, clamp lower circle
- release upper plate, set bearing of tangent
- define tangent by peg set at suitable distance
- set bearing of first point on curve and set peg at corresponding
  chainage
- continue setting out further points until table indicates need to
  move instrument
- move theodolite to last point set out
- set calculated bearing of origin (TTP) and align on TTP
- continue setting bearings of further points on curve.

Note: Bearings may be given as relative or whole circle bearings, but
the principles are the same.

If deflection angles are given rather than bearings, these are set
from the tangent at the origin.

An extract of a computer printout for a major road is given in
Appendix H. *p. 120*

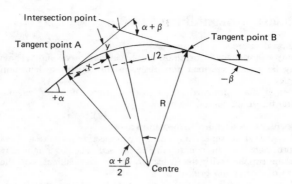

EXAMPLE:

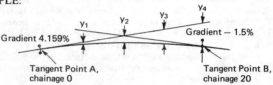

Intersection Point Level 94.717

$$R = \frac{20 \times 100}{(+4.159) - (-1.5)} = \underline{353.419 \text{ m}}$$

Level at Tan Pt. $A = 94.717 - \left(10.0 \times \frac{4.159}{100}\right) = \underline{94.301}$

Level at Tan Pt. $B = 94.717 - \left(10.0 \times \frac{1.5}{100}\right) = \underline{94.567}$

Select interval of 5 m along curve.

Change in level per 5 m along original gradient = $\underline{0.208 \text{ m}}$

$$y_1 = \frac{25}{2 \times 353.419} = 0.0354 \text{ m}$$

$$y_2 = 2^2 \times 0.0354$$
$$y_3 = 3^2 \times 0.0354$$
$$y_4 = 4^2 \times 0.0354$$

Roads: vertical circular curves

General equations and method

To set out vertical curves the following are needed:
- gradients (%) to be connected (α, $-\beta$) (Convention: positive indicates reduced level of road increases with chainage and vice versa.)
- distance between tangent points, L (or radius, R)
- reduced level (and coordinates) of tangent point or intersection point from which curve is to be set out.

Change of gradient $= \alpha - (-\beta) = \alpha + \beta$

For small angles, $\sin \simeq \tan = $ gradient (%) $\div 100$

Hence, $\dfrac{\alpha + \beta}{2 \times 100} = \dfrac{L/2}{R}$

and $R = \dfrac{100L}{\alpha + \beta}$ or $L = \dfrac{R(\alpha + \beta)}{100}$

Also $y = \dfrac{x^2}{2R}$ [from $x^2 = (R + y)^2 - R^2$, neglecting y^2]

To obtain height of points on curve:
- select convenient intervals for x ($\ngtr 10$ m)
- calculate levels for intervals along original gradient
- calculate corresponding values of y (*note*: y increases as square of interval i.e. in ratio $1:4:9:16:25$ etc.)
- deduct y to obtain height at distance x from A

Note: If height of intersection point is given, calculate heights along original gradient as shown opposite.

Worked example

Table of chainage and heights from calculated values:

CH. (m)	Point	Original grade: level change	Original grade: reduced level	y	Reduced level on centre-line
0	A	0	94.301	0	94.301
5		0.208	94.509	0.035	94.474
10	IP	0.416	94.717	0.141	94.576
15		0.624	94.925	0.318	94.607
20	B	0.832	95.133	0.566	94.567

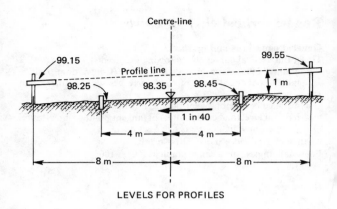

LEVELS FOR PROFILES

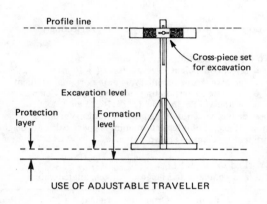

USE OF ADJUSTABLE TRAVELLER

80

Roads: levels

Temporary bench marks
- TBMs should be established and 'proved' *p. 25*

Cross-sections
Provide profiles generally at 20 m intervals and preferably at 10 m intervals.

Levels not shown on drawings must be calculated by the site engineer using what information there is on the drawings and from nearby levels and gradients. All such calculations must be shown on a site information sheet, and filed for reference purposes. *p. 7*

Note: It is usually possible to position centre-line and offset pegs to coincide with the level points shown at cross- and longitudinal sections

Profiles
For each section where profile is to be provided:
- determine channel levels from drawings or calculation
- decide offset of profiles from centre-line
- calculate required profile levels, taking into account offset distances and adding 1 m for height of traveller (*see illustration*)
- set up profiles *p. 29*
- mark reduced level and chainage on profile
- provide details of profiles on site information sheets
- check profiles daily for disturbance by sighting over three or more profiles.

Travellers
A convenient form of traveller is illustrated. The height of the cross piece can be adjusted from say, 1.5 m to 1 m as the construction proceeds.

Note: Excavation may be higher than the formation level if a protection layer is required to allow trafficking without damage to the formation.

After earthworks complete
After earthworks are complete, the need for accuracy is imperative because of the cost of materials and riding quality (Clause 702, DTp Specification *see* Bibliography). Set pins or kerbs to level for sub-base, base and surface construction. Alternatively, wire lines may be required to guide paving equipment. Check levels with the blacktop subcontractor. *p. 122*

Suggested colour code for profiles
A suggested colour code for profiles (and pegs) is shown on the back end-papers.

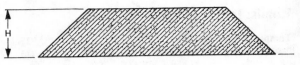

Section A-A

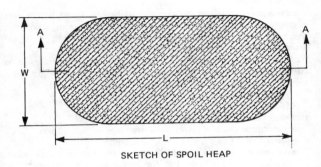

SKETCH OF SPOIL HEAP

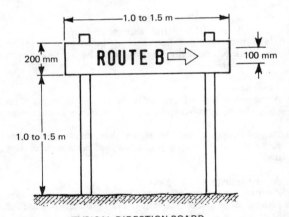

TYPICAL DIRECTION BOARD

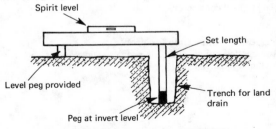

SETTING FORMATION DRAIN TO LEVEL

Roads: earthworks and drainage

Existing ground levels
Before site stripping starts:
- set up profiles at predetermined intervals   *p. 81*
- record existing ground levels at each cross-section
- check recorded levels against those on drawings
- agree recorded levels with the supervising authority.

Note: Recording existing ground levels can conveniently be combined with setting up profiles.

Spoil heaps
The site engineer should:
- set out the plan position with 2 m stakes painted white
- provide plan and profile of each spoil heap on an information sheet for the foreman/ganger concerned.

Haulage routes
The site engineer should:
- mark all haulage routes with direction boards that can be easily seen by the plant drivers, taking into account which public roads may not be used
- set lines and levels for routes that have to be constructed with fill to take heavy traffic.

Shoulder drainage
To control the levels of ditches and drains in shoulders:
- set up profiles as for sewers   *p. 45*
- provide profile and traveller details on a site information sheet for foreman/ganger.

Formation drains
Formation drains may be required under the road. The accuracy needed in positioning these drains is usually not great, and most of the setting out can be left to the foreman/ganger concerned.
   The site engineer should:
- provide foreman/ganger with drain layout on site information sheet
- establish pegs at suitable heights above drain levels
- make spot checks on line and level before backfilling.

Note: The ganger can transfer levels from peg to bottom of trench by means of levelling board and spirit level.

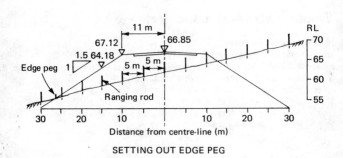

SETTING OUT EDGE PEG

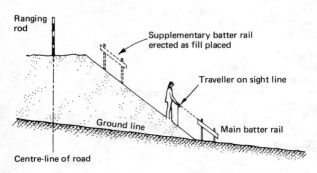

CONTROL OF SLOPE FROM BATTER RAIL

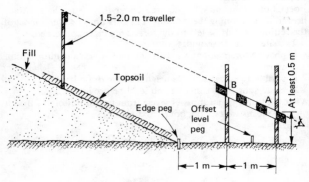

DETAIL OF BATTER RAIL

Roads: embankments

Setting-out edge pegs

In the office, prepare a table for each relevant chainage as in the example below (*illustrated opposite*). Calculate and enter the italicised slope line levels.

Distance from centre-line (m)	Ground level	Slope line level	Levels on batter stakes	Point
0	61.93			
5	60.97			
10	60.34			
15	59.28	*64.18*		
20	58.56	*60.85*		
25	57.21	*57.52*		
30	56.63	*54.18*		
25.5	57.15	57.18		Edge
26.0	57.09	56.85		
26.5	57.04	56.52	58.02	B
27.5	56.92	55.85	57.35	A

Distance up slope $= (25.5 - 11) \times \dfrac{\sqrt{(1.5^2 + 1^2)}}{1.5} = 17.43 \text{ m}$

Setting up batter rails

At relevant chainage on site:
- set ranging rods at 5-m intervals offset from centre-line peg
- measure and record ground levels at these intervals
- by inspection and/or calculation, find distance from centre-line where slope line intersects the ground line, i.e. the embankment edge
- mark the edge of the embankment with a peg
- set offset level peg at 1.5 m from edge of embankment
- drive two stakes 1 and 2 m from edge peg
- calculate and mark levels of point A and B for suitable traveller length (e.g. 1.5 m) and fix batter rail
- if embankment is more than 5 m high, set up batter rails on the embankment slope at height intervals of 5 m as fill proceeds
- record chainage and slope distance on batter rail
- set up ranging rods on the centre line to guide plant
- provide details of batter rails and traveller length on a site information sheet.

Controlling the work

See Roads: cuttings

p. 86

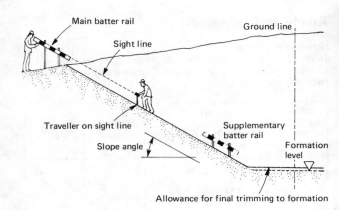

Main batter rail

Ground line

Sight line

Traveller on sight line

Supplementary batter rail

Slope angle

Formation level

Allowance for final trimming to formation

CONTROL OF SLOPE FROM BATTER RAIL

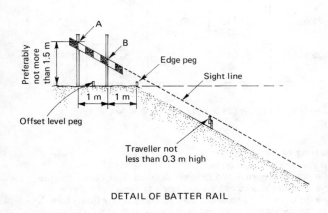

A

B

Preferably not more than 1.5 m

Edge peg

Sight line

1 m   1 m

Offset level peg

Traveller not less than 0.3 m high

DETAIL OF BATTER RAIL

Roads: cuttings

Setting out edge pegs

Follow a similar procedure as for embankments, calculating levels p. 85 on the slope line at intervals from the centre-line. Then, on site, find where the slope line intersects the ground line. Mark this point with an edge peg.

Setting up batter rails

- set offset level peg at 1.5 m from edge of cutting
- drive two profile stakes 1 and 2 m from edge peg
- calculate and mark levels of points A and B for a suitable traveller length (not less than 0.3 m) and fix batter rail
- if cutting is more than 5 m deep, set up batter rails on the cutting slope at depth intervals of 5 m as cut proceeds
- record chainage and distance down slope on batter rail
- set up ranging rods on the centre–line to guide plant
- provide details of batter rails and traveller length on a site information sheet.

Controlling the work (cuttings and embankments)

The site engineer should:

- check batter boards, profiles and travellers regularly
- reset ranging rods on centre–line as cut or fill proceeds
- check width of cut or fill either side of centre-line regularly
- when a *cut* is an agreed amount above formation level (typically 300 mm) set up standard road works profiles to control final trimming and road construction.
- when an *embankment* is nearly complete, set temporary profiles (allowing for settlement and trimming to formation)
- when agreed by supervising authority, set final profiles for trimming to formation and pavement construction
- check the final formation level.

Note: In deep cuttings, the ground may heave due to removal of overburden. It may be necessary to check that heave has ceased before setting up the roadworks profiles.

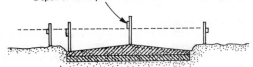

Depth of rail equals crossfall dimension

USING TRAVELLER RAIL TO CONTROL CROSSFALLS

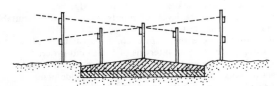

Section A: Equal crossfalls

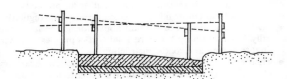

Section B: Developing super-elevation

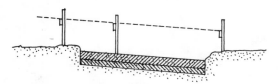

Section C: Superelevation fully developed

USING PROFILE RAILS TO CONTROL CROSSFALLS

Road: crossfalls

Each carriageway of a dual carriageway is usually constructed with a single crossfall but most single carriageways are built with crossfalls from the centre-line so that rain runs off to both sides. The crossfalls of the finished road are largely set by the cross-section of the formation. Thus it is important to shape the formation accurately.

Setting profiles
To be able to sight across the road, the profile boards must be parallel rather than perpendicular to the centre-lines. Profiles should be set at say, 10-m intervals.

Using traveller rail to control crossfall
For narrow or very minor roads, it may be sufficient to check the levels at the centre-line and road edges only and to check in between these levels by eye or with a straight-edge.
- make the depth of the traveller rail equal to the crossfall
- sight to the top of the traveller rail at the road edges
- sight to the underside of the traveller rail at the centre-line

Using profiles rail(s) to control crossfalls
More usually, it is necessary to check levels right across the section. In general, set two rails on each profile so that sight line from upper rail on one side to lower rail on the far side controls the far crossfall and vice versa (see Section A).

When developing super-elevation the two rails will become closer (see Section B), and it may be necessary to use two profiles side by side.

When the super-elevation is fully developed (see Section C), this provides the crossfall and each profile needs only a single rail.

Supervision
The site engineer should:
- give details of the profiles and travellers to the foreman/ganger concerned on a site information sheet
- check that the profiles and traveller are being correctly used.

Note: The traveller can conveniently be made adjustable as described previously. *p. 81*

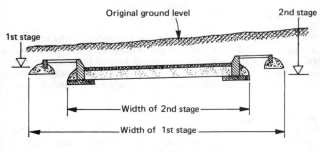

Original ground level

2nd stage

1st stage

Width of 2nd stage

Width of 1st stage

TWO-STAGE EXCAVATION

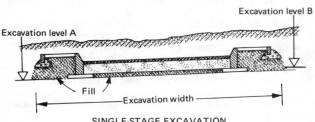

Excavation level A

Excavation level B

Fill

Excavation width

SINGLE-STAGE EXCAVATION

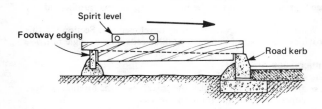

Spirit level

Footway edging

Road kerb

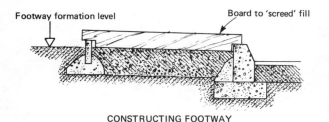

Footway formation level

Board to 'screed' fill

CONSTRUCTING FOOTWAY

Roads: footways and kerbs

When roads incorporate footways, excavation can be undertaken in two stages or a single stage. Site management will make this choice, depending on such factors as plant to be used, and costs of carting away excavated material and importing fill material.

Two-stage excavation
- set up profiles 1 m beyond each footway
- excavate over full width of road and footways to footway formation level
- reset profiles 1 m beyond each kerb line
- excavate over road width only to road formation level (boning along and across road to control crossfall)
- excavate as necessary along kerb line for kerb bed
- excavate for footway edging.

Note: This method is best suited to short lengths of road where heavy earthmoving plant is not appropriate.

Single-stage excavation
- set up profiles 1 m beyond each footway
- excavate over full width of road and footway to kerb foundation level (level *A*) or road formation level (level *B*) as preferred (the former choice may require importing additional fill to road formation level).

Note: This method is suitable where heavy earthmoving plant is justified by scale of the job.

Kerb and channel levels
The site engineer should:
- establish pegs at kerb level, 0.5 m behind face of kerb (intervals: 10 m on straights or moderate horizontal curves; 5 m on tight curves)
- if channel required, because there is no longitudinal fall, mark each 'valley' and 'summit' on face of kerb in waterproof yellow chalk. Kerblayer will use chalk line between valley and summit to lay concrete channel
- provide foreman/ganger with details on a site information sheet.

Constructing footway
Crossfall and width of footway can be controlled by using notched board and spirit level as shown. Fill between kerb and edging as necessary and 'screed' to formation level.

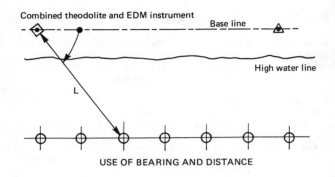

USE OF BEARING AND DISTANCE

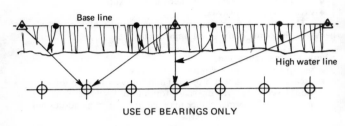

USE OF BEARINGS ONLY

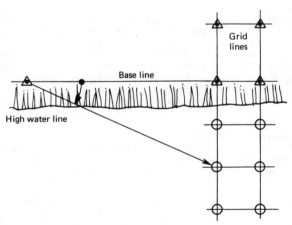

USE OF GRID LINES AND BEARINGS

Marine and river works: bearing piles

Bearing piles are taken to include piles capable of resisting lateral loads (e.g. mooring loads) as well vertical loads. Such piles may be driven as raking piles.

Method of installation

It is usual to drive temporary piles from a moored floating platform and then to erect a temporary staging from which to pitch and drive the permanent piles. Each pile is held in a 'gate' (or guide) during driving to achieve the specified tolerances for location and rake; of necessity these tolerances are commonly wider than for land-based piling.

The difficulties

The major difficulties in setting out are measuring distances over water and providing stable reference points in deep water. A (temporary) reference pile can incorporate a sighting mark but check for oscillation of the pile, particularly in fast currents.

Check that sight lines to or from water level will be unobstructed whatever the tide state.

Communication can be difficult; a radio link is advisable.

Locating piles by bearing and distance

Probably the most convenient method of locating piles over water is to use a theodolite and EDM equipment combined, so that one engineer can indicate both bearing and distance.

Locating piles by bearings only

For lines of piles roughly parallel to the high water line (or river bank) an alternative method may be to use bearings only from an appropriate pair of reference points. The disadvantage of this method is that 'three-way' communication is required.

Locating piles with grid lines and bearings

For piles in lines roughly perpendicular to the shore line (or river bank), where practicable:

- establish reference points defining these grid lines
- set up reference point some distance along the shore and calculate bearing of each pile from this point
- for each pile use two theodolites to set out the grid line and the intersecting bearing.

Note: If piles are constructed from the shoreline, tape along grid line on the temporary staging used for piling.

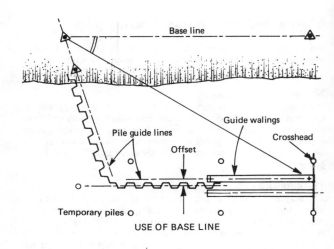

USE OF BASE LINE

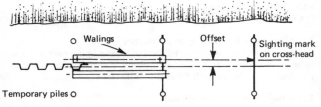

USE OF SIGHTING MARK

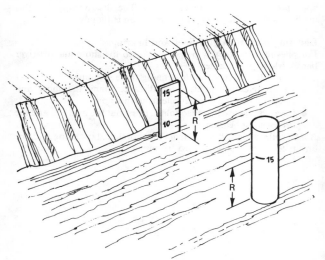

USE OF STAFF GAUGE

94

Marine and river works: sheet piles

Method of installation
As on land, sheet piles are driven in 'panels', using a pair of walings as a guide. Pairs of temporary piles are driven at 'panel' intervals either side of the sheet pile centre line. A crosshead, located on a pair of temporary piles, supports the 'leading' end of the walings. The 'trailing' end is supported on the end pile of the panel of sheet piles previously driven. Thus it is necessary to check the line of the leading end of the waling before each panel is driven.
Note: Temporary piles are often later used to support falsework.

Line of piles at angle to shore
For sheet piles at an angle to shore line/river bank:
- set up theodolite on line offset to centre-line of piles
- fix offset targets at each end of waling
- sight along offset centre-line and align both ends of waling
- after driving first panel of piles, set trailing end of waling against last pile driven and align leading end.

Line of piles roughly parallel to shore
Set up a base line along the shore/river bank and use bearing and distance or two bearings to set up the walings as described previously. Alternatively, sighting marks could be set up on temporary piles. *p. 93*

Driving piles to level
To drive piles to required level with fair accuracy in reasonably calm water:
- set up staff gauge to be read from piling rig (staff gauge indicates tide/river level relative to OD)
- mark suitable OD level on pile *above* HWL
- drive pile until mark on pile and same OD level on staff gauge are same distance above water level.

Underwater concreting
Having constructed a sheet pile cofferdam, it may be necessary to concrete a base underwater. If formwork is required, it must be positioned by divers. Within reasonable limits of accuracy this can be achieved as follows:
- set out each locating point from the shore stations
- fix plywood sheets (or timbers) to the upper walings
- drill small holes through ply and suspend plumb bobs from piano wires passed through the holes
- when divers have located formwork by plumb bobs, use weighted tape to define required concreting level.

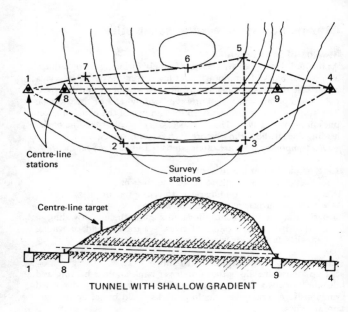

Centre-line stations

Survey stations

Centre-line target

1   8   9   4

TUNNEL WITH SHALLOW GRADIENT

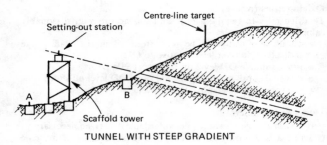

Setting-out station

Centre-line target

Scaffold tower

A   B

TUNNEL WITH STEEP GRADIENT

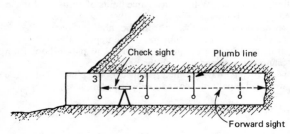

Check sight

Plumb line

3   2   1

Forward sight

PROJECTING CENTRE-LINE FORWARD

Tunnelling: from ground level

Tunnel with shallow gradient
Assuming that points 1 and 4 define the centre-line:
- determine the whole circle bearing (WCB) of 4 from 1 by carrying out a closed traverse
- set up theodolite on station 1, set WCB of survey station 2 and align telescope on that station
- set out centre-line station 8 for portal construction
- set out centre-line target above and clear of future tunnel works to replace station 8 when portal complete
- when section of tunnel constructed transfer centre line on to roof at three well-spaced sections.

Tunnel with steep gradient
Assuming that centre-line stations A and B have been set out:
- set up centre-line target above tunnel works
- erect heavily braced scaffold tower between A and B
- set out station C on top of scaffold tower
- with theodolite on station C, align on centre–line target, and set vertical angle to gradient of tunnel to provide line parallel but offset vertically to centre line
- check vertical offset by levelling at tunnel portal

Note: Line and gradient can subsequently be conveniently controlled by laser but check frequently with theodolite.

Projecting centre line forward
- suspend plumb lines from three centre-line marks nearest tunnel face (*see illustration*)
- set up theodolite between plumb lines 2 and 3
- sight towards face and traverse telescope until offsets of plumb lines 1 and 2 are equal
- traverse through 180° to check offset of plumb line 3
- if offsets agree, traverse again through 180° and set one or more new (offset) centre-marks on tunnel roof
- if offsets disagree, recheck suspect centre-marks from proven centre-marks further from face.

Note: A laser can also conveniently be used. *p. 39*

Need for accuracy
- Use a theodolite reading to 1 second and average a number of readings.

Note: An error of 20 seconds is equivalent to an alignment error of about 100 mm per km.

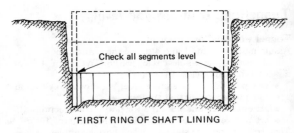

Check all segments level

'FIRST' RING OF SHAFT LINING

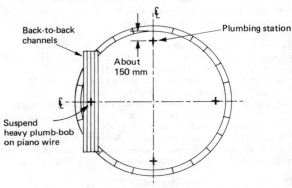

Back-to-back channels

Plumbing station

About 150 mm

Suspend heavy plumb-bob on piano wire

PLUMBING SHAFT

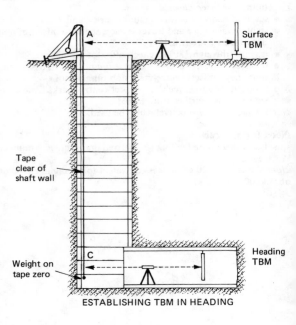

A

Surface TBM

Tape clear of shaft wall

C

Weight on tape zero

Heading TBM

ESTABLISHING TBM IN HEADING

98

Tunnelling: constructing shafts

Setting out first ring

Most shafts are circular and lined with precast concrete or cast iron bolted segments. The first ring constructed is usually the third ring down.

When 'first' ring has been set up with bolts hand-tight only:
- re-establish centre-point
- check concentricity of the segments
- check uniformity of level of all segments

The bolts can then be tightened, the rings above constructed and a concrete 'collar' placed for stability. If these first rings are set and maintained true, subsequent adjustments for line and level need only be minimal.

Plumbing down shaft

Check verticality of shaft as follows:
- establish four plumbing stations about 150 mm from shaft wall at ends of two orthogonal centre lines
- suspend plumb lines from these stations (damp movement of bobs by suspending in oil or water)
- in deep shafts where visibility is poor, check that plumb line is clear of shaft well by dropping 'snap' washer down plumb line
- check verticality by measuring offset at 'first' ring and last ring constructed

Alternatively, a suitable optical plumbing instrument or laser may be set up over each plumbing station.

p. 31

Establishing TBM in heading
- suspend weighted tape from timber frame at top of shaft
- set up level at surface, sight first on surface TBM, determine collimation level and read tape at A
- set up level in heading, read tape at B, and calculate collimation level
- determine reduced level of TBM on roof of heading

Note: Weight on tape should be calibration load less half the self-weight of the suspended portion of the tape. Where possible, use full depth tape. If this is not practicable, fix suspension points at suitable intervals down shaft and tape between these. For very deep shafts, electronic distance-measurement (EDM) may be appropriate.

Warning: Shot-fired pins are convenient for TBM and other levelling points in rock tunnels but check safety aspects, especially methane risk.

p. 1

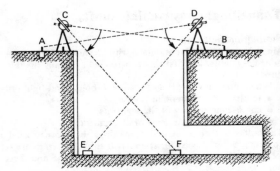

TRANSFERRING LINE DOWN SHALLOW SHAFT

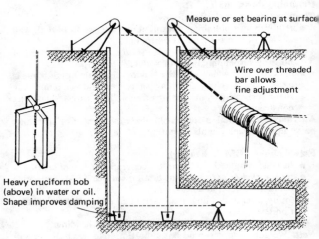

Measure or set bearing at surface

Wire over threaded bar allows fine adjustment

Heavy cruciform bob (above) in water or oil. Shape improves damping

PLUMBING LINE DOWN SHAFT

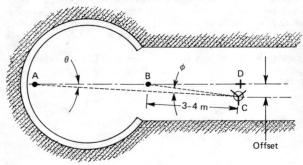

θ ϕ

3–4 m

Offset

WEISBACH TRIANGLE

Tunnelling: from shafts

Transferring line down shallow shaft
With A and B defining the required heading direction:
- set up subsidiary stations C and D close to shaft
- from C and D, set out F and E respectively

Note: Check accuracy by making a number of observations.

Plumbing line down shaft
Two plumb lines are suspended down the shaft, as far apart as practicable. The two lines may define:
- given bearing or direction

or
- random bearing which is measured at the surface

'Picking up' bearing
Whether at the surface or underground, the bearing defined by the plumb lines has to be 'picked up' by using a 1-second theodolite and the Weisbach triangle (see below). Underground, illuminate each plumb line (piano wire) with a lamp.

Use of the Weisbach triangle
- set up theodolite at C, about 3–4 m from B and offset by say 25–50 mm from the line defined by AB
- measure the angle subtended by A and B, ϕ (average of a number of readings on both faces)
- measure the lengths AB and BC
- calculate $\theta = \phi \cdot \dfrac{BC}{AB}$, where θ and ϕ are in seconds
- calculate offset, $CD = \dfrac{\theta(AB + BC)}{57.2958 \times 3600}$
- align on A and traverse through $180° - \theta$ (clockwise as illustrated) to sight along line by calculated offset
- mark centre-line by taping from offset line (appropriate for small tunnels or tunnels with bends)

or
- mark further line offset from centre-line by convenient amount up to, say, 1 m (appropriate for large tunnel where centre-line may be obscured by ventilating duct).

Gyro-theodolite
A gyro-theodolite can provide an accurate bearing where other methods would be impossible. Setting up a gyro-theodolite is a specialist operation. *See* Bibliography.

p. 122

Controlling squareness
In large tunnels where a primary lining is used, the control of line is related to the control of the 'squareness' of the lining. To check squareness:
- set up theodolite on (offset) centre-line
- turn through 90° (left and right)
- rotate telescope to 45° above and below horizontal and mark lining
- tape from the four marks to lining at face.

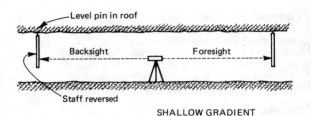

Level pin in roof

Backsight

Foresight

Staff reversed

SHALLOW GRADIENT

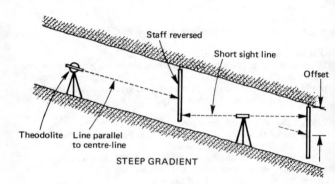

Staff reversed

Short sight line

Offset

Theodolite

Line parallel to centre-line

STEEP GRADIENT

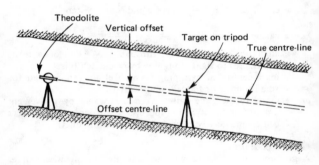

Theodolite

Vertical offset

Target on tripod

True centre-line

Offset centre-line

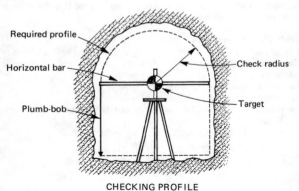

Required profile

Check radius

Horizontal bar

Plumb-bob

Target

CHECKING PROFILE

Tunnelling: gradients, profiles and length

Controlling gradients
Gradients may be controlled by:
- theodolite set to required gradient (convenient when theodolite is clear of operations) p. 97
- profiles fixed to roof (moderately convenient for excavation by drilling and blasting or hand digging)
- laser aligned to gradient, established by conventional methods (convenient for all methods of excavation) p. 39

Checking levels on shallow gradient
Check gradient and reduced levels at regular intervals:
- establish underground TBM and level points in roof p. 99
- level from underground TBM using staff reversed

Note: TBM and level pins may have to be offset from centre-line to avoid ventilation ducts, etc.

Checking levels on steep gradient
Standard tunnel levelling procedure may be used but note that short sight lines tend to reduce precision. Alternatively, check with theodolite:
- set vertical circle on theodolite to gradient
- using reversed staff, check offset of known roof level
- compare offsets of pins further forward.

Profile of cross-section
The site engineer may have to check selected cross-sections for measurement purposes or because ground movement is suspected. A suitable technique is as follows:
- align theodolite on offset centre-line and to gradient (for convenience, directly below a proven level)
- measure vertical offset from centre-line
- set up tripod and target at selected cross-section
- align target with offset centre-line
- adjust target height by offset distance to establish centre-point of section
- check profile dimensions by taping or using a trammel

Note: A horizontal cross bar may be necessary if the cross-section is non-circular (*see* illustration).

Controlling length
Controlling length may be critical if bends or other features must be located precisely. An invar band, rather than a steel tape, may be necessary together with catenary taping techniques, using firm supports with well-defined datums p. 21

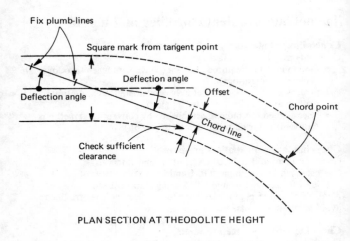

PLAN SECTION AT THEODOLITE HEIGHT

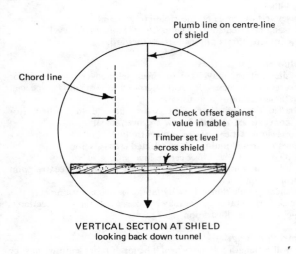

VERTICAL SECTION AT SHIELD
looking back down tunnel

Tunnelling: around curves

Procedure in the office
- on plan of tunnel, determine suitable chord(s) that will not be too close to tunnel wall along line of sight and calculate deflection angle
- on plan, mark off ring width on inside radius of tunnel
- calculate offset from chord to centre of each ring
- calculate required 'lead' per ring on outside radius (lead arises from greater arc length on outside of curves).

Procedure in the tunnel
- with theodolite at tangent point, sight back down tunnel on centre-mark and traverse through deflection angle
- set up two plumb lines on chord line in straight portion of tunnel
- set out square marks from tangent point
- as excavation proceeds extend chord line into curved portion and set up plumb lines
- check offsets from chord to centre of ring/shield
- when shield is clear of first chord point, set out this point accurately by bearing and distance
- set up on first chord point, sight on tangent point and traverse to set up second chord line
- set out square marks at first chord point
- check any deviation of the tunnel at the first chord point and distribute the error to bring tunnel back onto true centre by or before the second chord point
- repeat establishment of new chords as necessary.

Site information sheet
A site information sheet for the foreman/ganger (copy to RE) should be prepared by the site engineer and include a table with headings as below:

| Ring No. | Offset (mm) | Lead from mark | |
		Ring No. marked	Distance (m)

ICE Conditions of Contract: Fifth Edition, Clause 17

The Contractor shall be responsible for the true and proper setting-out of the Works and for the correctness of the position levels dimensions and alignment of all parts of the Works and for the provision of all necessary instruments appliances and labour in connection therewith.

If at any time during the progress of the Works any error shall appear or arise in the position levels dimensions or alignment of any part of the Works the Contractor on being required so to do by the Engineer shall at his own cost rectify such error to the satisfaction of the Engineer unless such error is based on incorrect data supplied in writing by the Engineer or the Engineer's Representative in which case the cost of rectifying the same shall be borne by the Employer. The checking of any setting-out or of any line or level by the Engineer or the Engineer's Representatives shall not in any way relieve the Contractor of his responsibility for the correctness thereof and the Contractor shall carefully protect and preserve all bench-marks sight rails pegs and other things used in setting out the Works.

(Reproduced by permission of the Institution of Civil Engineers)

JCT Standard Form of Building Contract (1980 Edition with revisions to Nov. 1986): Conditions, Clauses 2 (extracts) and 7

2.1 The Contractor shall upon and subject to the Conditions carry out and complete the Works shown upon the Contract Drawings and described by or referred to in the Contract Bills and in the Articles of Agreement, the Conditions and the Appendix . . .

2.3 If the Contractor shall find any discrepancy in or divergence between the description in the Contract Bills and any one or more of the following documents, . . . namely,

    2.3.1 the Contract Drawings

    2.3.3 any instruction issued by the Architect . . .

    2.3.4 any drawings issued by the Architect . . .

he shall immediately give to the Architect a written notice specifying the discrepancy or divergence, and the Architect shall issue instructions in regard thereto.

7. The Architect shall determine any levels which may be required for the execution of the Works and shall provide the Contractor by way of accurately dimensioned drawings with such information as shall enable the Contractor to set out the Works at ground level. Unless the Architect shall otherwise instruct, in which case effect shall be given to the Architect's instructions in the calculation of the Ascertained Final Sum, the Contractor shall be responsible for and shall entirely at his own cost amend any errors arising from his own inaccurate setting out.

Note: In Local Authorities' versions, for 'Architect' read 'Architect/ Supervising Officer'.

(Reproduced by permission of RIBA Publications Ltd)

Appendix A.
Contractual aspects

Responsibilities

The responsibilities of the Engineer, Architect or Supervising Officer and the Contractor, in respect of setting out the Works, are normally laid down in the Conditions of Contract. Most Works are carried under the ICE Conditions of Contract or one of the variants the JCT Standard Form of Building Contract and the relevant clauses of these two documents are reproduced opposite. The salient points to note are:

- the Contractor bears the responsibility for setting out the Works
- the Contractor generally bears the cost of rectifying errors arising from incorrect setting out
- the Contractor must not rely on any checking of the setting out by the Engineer or the Architect or their representatives
- the Engineer, Architect or Supervising Officer must provide, *in writing*, essential data required by the Contractor for setting out the Works
- the Contractor may engage a specialist firm to carry out the main setting-out operations but should check the results independently
- similarly, the Contractor may agree that a Sub-Contractor shall set out his part of the Works but should check the results independently and at frequent intervals.

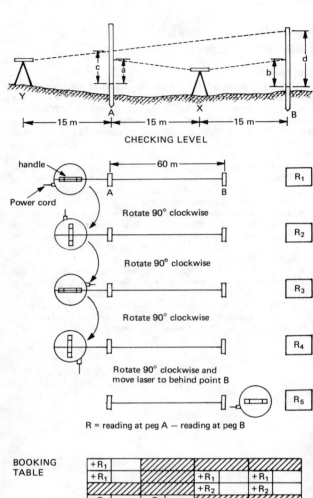

R = reading at peg A — reading at peg B

BOOKING TABLE								
+R₁							
+R₁				+R₁		+R₁	
				+R₂		+R₂	
+R₃		−R₃					
				−R₄		+R₄	
−R₅		−R₅		−R₅		−R₅	
Column totals							

Each column total to be < 6 mm for accuracy

CHECKING ROTATING LASER

Appendix B.
Care and checking of equipment

Care and maintenance
Site engineers are responsible for the day-to-day care and mainten-
ance of instruments.

All setting-out instruments must be packed away carefully when
not in use, using silica-gel bags to ensure dry conditions. Clean
lenses with recommended tissue or dusting brush. Instruments (and
tripods) are expensive and must not be left on site unattended—
even for short periods. Damage or total destruction is often caused
by mobile plant.

When carrying instruments by road, strap case in vehicle.

Adjusting instruments
Most companies prefer that instruments be adjusted by the manufacturer or agent. If adjustment by the site engineer is permitted,
follow manufacturers' instructions strictly.

Checking level
A level should be checked at least weekly (two-peg test):
- from X, take readings a and b
- apparent difference in level of A from $B = a - b$
- from Y, take readings c and d
- if $[(a - b) - (c - d)]$ is more than say, 4 mm, arrange for
  adjustment.

Checking theodolite
Plate bubble: Set up the theodolite. Turn theodolite until the bubble
tube is parallel with two footscrews, and level the bubble. Turn
theodolite through 90°, level bubble using third footscrew. Turn
back through 90°, relevel bubble if necessary. Then turn through
180°, the bubble should still be central if correctly adjusted.
Optical plummet: Set up and level the theodolite over point defined
by a cross. Turn the theodolite through 180°, at the same time
observing the point. The point should remain central at all times.
Horizontal circle: Set up and level the theodolite. Sighting a well-
defined distant point, read the horizontal angle scale (H_1). Transit
the telescope and re-sight the point on the opposite face, read the
horizontal angle scale (H_2). The difference between H_1 and H_2
should be 180°.
Vertical circle: Set up and level the theodolite. Sighting a well-
defined elevated point, read the vertical angle scale (V_1). Transit the
telescope and re-sight the point on the opposite face, read the
vertical angle scale (V_1). (V_1) and (V_2) should be equal.
Trunnion axis: Set up and level the theodolite. Sight a well defined
high-level point with the centre of the cross-hairs. Depress the
telescope and read a tape set close to the theodolite (T_1). Repeat this
on the opposite face and read the tape again (T_2). The readings (T_1)
and (T_2) should be equal.

Checking rotating laser
Carry out 'two-peg' test as illustrated opposite.

Checking EDM equipment
Apart from the simple check given earlier, checking of EDM p. 41
equipment should be carried out by experienced operators.

Local scale factor for any part of England, Scotland or Wales

National Grid Easting (km)		Scale Factor F.
400	400	0.99960
390	410	60
380	420	61
370	430	61
360	440	62
350	450	63
340	460	65
330	470	66
320	480	68
310	490	70
300	500	72
290	510	75
280	520	78
270	530	81
260	540	84
250	550	88
240	560	92
230	570	0.99996
220	580	1.00000
210	590	04
200	600	09
190	610	14
180	620	20
170	630	25
160	640	31
150	650	37
140	660	43
130	670	1.00050

Appendix C.
National grid, bench marks and ground distances

National grid
All Ordnance Survey (OS) plans for England, Scotland and Wales are constructed on the National Grid. The grid is linked to the primary triangulation of Great Britain. For further details *see* Bibliography.

p. 122

Triangulation stations are usually specially erected pillars, marks on church towers, water tanks, buried concrete blocks and church spires.

Note: The description and National Grid co-ordinates of triangulation stations may be obtained for a fee from the Ordnance Survey (see address below).

Ordnance Bench Marks
A network of Ordnance Bench Marks covers the area of the National Grid. Some are incorporated into triangulation stations but most are built or marked on substantial buildings or structures.

A list of local bench marks for a specified kilometre square (e.g. TF 7934) may be obtained for a fee from the Ordnance Survey (see address below). For each bench mark, the list provides: description (building or structure incorporating mark), National Grid 10-m reference, altitude, height above ground and date of levelling.

Ground distances
The distance calculated from any two OS triangulation stations is the *projection distance* and differs from the *ground distance* (i.e. the true horizontal distance). To convert projection distance to ground distance, it is necessary to use a local scale factor.

$$\text{Thus ground distance} = \frac{\text{projection distance}}{\text{local scale factor}}$$

Example:

	National Grid Co-ordinates			
OS Station *B*	E.	482841.570	N.	263542.210
OS Station *A*	E.	482671.950	N.	363433.540
Difference		169.620		108.670

Projection distance $= \sqrt{(169.620^2 + 108.670^2)} = 201.445$ m

Ground distance $= \dfrac{201.445}{0.99968}$ $= \underline{201.510 \text{ m}}$

For details of O.S. triangulation stations and bench marks, contact:

> Survey Services
> Ordnance Survey
> Romsey Road
> Southampton SO9 4DH
> Tel. (0703) 792518 and 792519

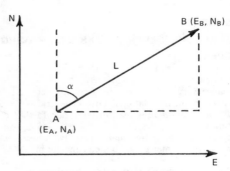

Rectangular and polar coordinates

Rectangular to polar conversion

| Most calculators | | RPN calculators (eg HP) | |
Keystrokes	Display	Keystrokes	Display
dN (±) $\boxed{R \rightarrow P}$ dE (±) $\boxed{=}$ $\boxed{x \rightleftharpoons y}$ $\boxed{\rightarrow DMS}$	dN (±) dE (±) L α (decimal) α (deg min sec)	dE (±) $\boxed{ENTER}$ dN (±) $\boxed{R \rightarrow P}$ $\boxed{x \rightleftharpoons y}$ $\boxed{DMS \rightarrow}$	 dE (±) dN (±) L α (decimal) α (deg min sec)

Polar to rectangular conversion

| Most calculators | | RPN calculators (eg HP) | |
Keystrokes	Display	Keystrokes	Display
L $\boxed{P \rightarrow R}$ α (WCB, dms) $\boxed{DMS \rightarrow}$ $\boxed{=}$ $\boxed{x \rightleftharpoons y}$	L α (dms) α (decimal) dN dE	α (WCB, dms) $\boxed{DMS \rightarrow}$ $\boxed{ENTER}$ L $\boxed{P \rightarrow R}$ $\boxed{x \rightleftharpoons y}$	α (dms) α (decimal) α (decimal) L dN dE

Notes:
1. The sequence of keystrokes only is indicated. Actual keystrokes vary with calculator model and make. Some functions, for example, may require two keystrokes.
2. Partial coordinates dE and dN must be input as positive or negative.
3. If output value for α is negative, add 360° to obtain WCB.
4. Check that correct angular mode (degrees) is set.

Appendix D
Converting from rectangular to polar co-ordinates

In the figure, the rectangular coordinates of A and B are (E_A, N_A) and (E_B, N_B) in relation to the grid origin. It is however, frequently more convenient to use the polar coordinates of one point in relation to another point which is not the grid origin. For example, the polar coordinates of B, with respect to A, are (L, α), where α is the whole circle bearing (WCB) and has any value from 0° to 360°.

Polar coordinates are derived from partial coordinates.

The partial coordinates of B with respect to A are $(E_B - E_A, N_B - N_A)$ or (dE, dN).
This order of subtraction gives the correct algebraic values.

It follows that
$$L = \sqrt{(dE^2 + dN^2)}$$
$$\alpha = \sin^{-1} dE/L$$
$$\text{or } \alpha = \cos^{-1} dN/L$$

The algebraic values of dE and dN show in which quadrant α lies.

Use of scientific calculators

Most scientific calculators have keys to convert rectangular to polar coordinates and vice versa. The general sequence of keystrokes for most calculators and for RPN (Reverse Polish Notation) calculators are indicated opposite. The actual keystrokes will vary with calculator make and model. The site engineer is advised to use the examples below to check the sequence and keys to be used and to record these on p. 10 as an aide-memoire.

Examples
1. Rectangular
coordinates:

	B	514.62E		438.21N
	A	428.96E		214.82N
Partial coordinates:	dE	85.76	dN	223.39

Polar coordinates of B (w.r.t. A): $L = 239.29$, $\alpha = 21° 00' 07''$

2. Rectangular
coordinates:

	C	311.09E		102.68N
	A	428.96E		214.82N
Partial coordinates:	dE	− 117.87	dN	− 112.14

Polar coordinates of C (w.r.t. A): $L = 162.69$, $\alpha = 226° 25' 37''$

3. Polar coordinates of E (w.r.t. D): $L = 387.92$, $\alpha = 147° 19' 03''$
   Partial coordinates: $dE = 210.64$, $dN = - 325.75$

4. Polar coordinates of G (w.r.t. F): $L = 204.31$, $\alpha = 304° 26' 42''$
   Partial coordinates: $dE = - 169.07$, $dN = 114.71$

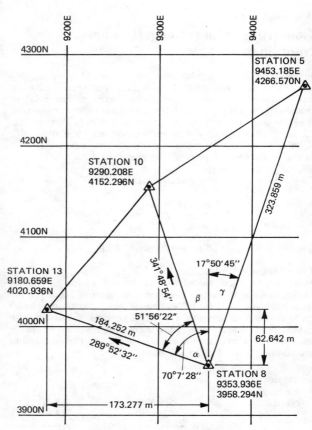

ANGLES AND DISTANCES CALCULATED
FROM GIVEN CO-ORDINATES

Appendix E
Road layout: proving setting-out stations

The procedure for proving the coordinates of setting-out stations is *p. 15*
given in the main text. An example is given below using the road
layout illustrated previously. *p. 14*

Calculate ground distance
From coordinates of stations 8 and 13 for example:

Station 13:	9180.659E	4020.936N
Station 8:	9353.936E	3958.294N
Partial coordinates 8 to 13:	− 173.277	+ 62.642

Calculated ground distance 8 to 13 = $\sqrt{(173.277^2 + 62.642^2)}$
= 184.252 m

Check ground distance
Assume that measured and calculated ground distances between
stations 8 and 13 agree within 20 mm.
  Therefore Stations 8 and 13 may define base line.

Calculate relative bearings
Referring to figure opposite, for stations 8 and 13:

$$\tan \alpha = \frac{\text{difference in eastings}}{\text{difference in northings}} = \frac{173.277}{62.642}, \; \alpha = 70° \, 7' \, 28''$$

Therefore, whole circle bearing (WCB) = 289° 52′ 32″
Repeat for Stations 10 and 8

Station 10:	9290.208E	4152.296N
Station 8:	9353.936E	3958.294N
Partial coordinates 8 to 10:	− 63.728	+ 194.002

$$\tan \beta = \frac{63.728}{194.002}, \; \beta = 18° \, 11' \, 6'', \; \text{WCB} = 341° \, 48' \, 54''$$

Repeat for Stations 5 and 8, γ = WCB = 17° 50′ 45″

Note: The calculation of ground distance and bearing is facilitated
by the use of polar coordinates on a suitable calculator but the
above method would provide a suitable cross-check.

Check relative bearings
- Set up theodolite on station 8
- Set upper plate to WCB for station 13 from station 8
- Align theodolite on station 13 and lock lower plate
- Release upper plate and observe WCB's when aligned on
  stations 10 and 5 in turn
- Repeat procedure on opposite face.

Check that the error between mean observed and calculated WCBs
is acceptable.

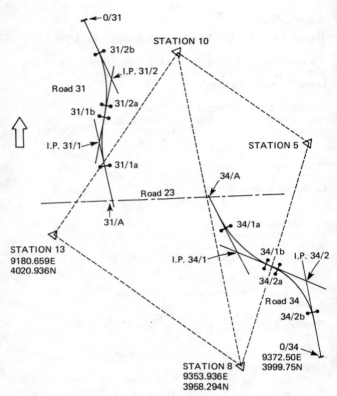

COORDINATE LAYOUT (some coordinates omitted)

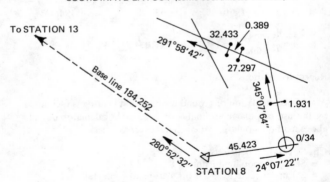

EXTRACT OF LAYOUT OF BEARINGS AND LENGTHS

Appendix F
Road layout: setting-out main points

When the setting-out stations have been 'proved' on site, the site engineer can calculate the bearings and distances of the main points and centre-lines of roads 34 and 31 in the layout. This is facilitated by drawing skeleton layouts as shown opposite. The top layout shows the centre-lines and (some) coordinates of centre points and intersection points. Tangent points are 34/1a, 34/1b, etc. The lower layout is a basis for recording calculated bearings and distances for use on site.

The calculations should be done in the office.

Starting from Station 8, to fix the base point of the hammer head on road 34 (0/34), coordinates are converted into bearings and distances thus:

0/34:	9372.500E	3999.750N
Station 8:	9353.936E	3958.294N

Partial coordinates

| 8 to 0/34: | + 18.564 | + 41.456 |

$$\alpha = WCB = \tan^{-1}\frac{18.564}{41.456} = 24° \ 7' \ 22''$$

The distance from 8 to 0/34 $= \sqrt{(18.564^2 + 41.456^2)}$

$$= \underline{45.423 \ m}$$

Note: The calculation of ground and bearings is facilitated by the use of polar coordinates on a suitable calculator but the above *p. 113* method would provide a suitable cross-check.

Similar calculations are repeated along the main line of roads 34 and 31. To ensure the setting out 'closes', check bearings and distances from alternative stations are added.

The site engineer can now start setting out the main points on site using colour coded pegs as described previously. Setting up the theodolite over station 8, fix bearing 289° 52' 32" and sight on station 13. Then reading bearing 24° 07' 22", measure 45.42 m, taking into account slope. This will determine point 0/34. This sequence is repeated along the main lines of roads 34 and 31 starting at 0/34. Check bearings are taken from station 13 and station 10 to 0/31.

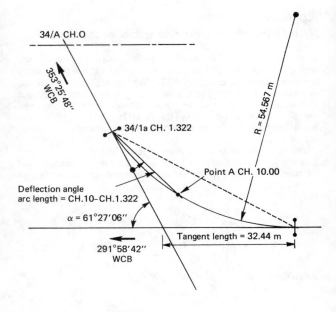

Setting out centre line of curve

Station	Chainage of Chord PT m	Arc length m	Deflection angle Deg. Min. Sec.	W.C.B. reading Deg. Min. Sec.	Chord length m
(1)	(2)	(3)	(4)	(5)	(6)
34/1a	1·322			173 25 48	
'A'	10·000	3·678	04 33 22	168 52 26	8·669
	20·000	10·000	05 15 00	163 37 26	9·986
	30·000	10·000	05 15 00	158 22 26	9·986
	40·000	10·000	05 15 00	153 07 26	9·986
	50·000	10·000	05 15 00	147 52 26	9·986
34/1b	59·847	9·847	05 10 11	142 42 15	9·834
		$\alpha/2$ = (30 43 33)		(111 58 42)	
		α = 61 27 06			

Appendix G
Road layout: setting-out horizontal curves

This example is based on one of the curves of the estate road layout
given previously for which the coordinates and bearings of the
skeleton have been worked out (Appendix F).

p. 14
p.117

Given, in addition to the tangent bearings:

$R = 54.567$ m, tangent length = 32.44 m

Deflection angle of point A and bearing of tangents
Arc length 34/1a to A = CH10.00 − CH 1.322

$\qquad = 8.678$ m = $R \times 2\alpha$ (α in rad)

Deflection angle α (deg) = 4° 33′ 22″

Whole circle bearing (WCB) of A from 34/1a

= 173° 25′ 48″ − 4° 33′ 22″ = 168° 52′ 26′

Chord length 34/1a to $A = 2R \sin \alpha = \underline{8.669 \text{ m}}$

Repeat for points 2, 3, 4, 5 and 34/1b as shown in table.

Using proforma shown opposite calculate all the deflection
angles at all the chosen points along the arc (col. 4). The sum of
these deflections should equal $\alpha/2$. The WCB is then worked out to
each of the chainages.

Assuming theodolite is set up at 34/1a,
final bearing to 34/1b = 142° 42′ 15″
To check:

142° 42′ 15″ = bearing of the tangent straight
IP 34/1 to IP 34/2 = 291° 58′ 42″ − 180° = 111° 58′ 42″
Final bearing = 142° 42′ 15″ − sum of deflections

$\qquad = 142° 42′ 15″ − 30° 43′ 33″$

$\qquad = 111° 58′ 42″$

Finally calculate chord lengths (col. 6)

Appendix K: Extract of computer printout for major road

APPENDIX K: EXTRACT OF COMPUTER PRINTOUT FOR MAJOR ROAD

POINT	-----X-----	-----Y-----	-----Z-----	-CHAINAGE-	-DEFLN ANG-			DEFLN DIST	CHORD DIST	
STB9	539546.913	211056.573	46.151							REFERENCE STATION
										POINT 285 TO STATION STB9
										BEARING 313 50 36.5
										DISTANCE 24.042
285	539564.253	211039.919	39.053	2800.000	0	0	0.0	0.000	0.000	
286	539571.661	211046.637	39.317	2810.000	93	57	15.2	10.000	10.000	
287	539579.069	211053.354	39.581	2820.000	93	57	15.2	20.000	10.000	
294	539630.923	211100.377	41.428	2890.000	93	57	15.2	90.000	10.000	
295	539638.331	211107.094	41.692	2900.000	93	57	15.2	100.000	10.000	
MOVE INSTRUMENT POINT 295										POINT 295 TO POINT 0285
										BEARING 227 47 51.7
										DISTANCE 100.000
296	539645.739	211113.812	41.956	2910.000	180	0	0.0	10.000	10.000	
297	539653.146	211120.529	42.219	2920.000	180	0	0.0	20.000	10.000	
305	539707.518	211169.835	44.156	2993.399	180	0	0.0	93.399	3.399	
306	539712.409	211174.269	44.330	3000.000	180	0	0.9	100.000	6.601	

MOVE INSTRUMENT POINT 306

	Y	X	Z	°	′	″			
307	539719.821	211180.982	44.591	180	2	7.8	3010.000		10.000
308	539727.242	211187.684	44.851	180	4	37.2	3020.000		20.000
309	539734.680	211194.369	45.109	180	8	6.6	3030.000		30.000
315	539779.947	211233.745	46.619	180	50	3.0	3090.000	89.994	10.000
316	539783.990	211237.106	46.748	180	55	26.5	3095.258	95.249	5.258
317	539787.650	211240.122	46.864	181	0	32.1	3100.000	99.989	4.742

POINT 306 TO POINT 0295
BEARING 227 47 52.5
DISTANCE 100.000

MOVE INSTRUMENT POINT 317

	Y	X	Z	°	′	″			
318	539795.409	211246.430	47.107	182	4	51.0	3110.000		10.000
319	539803.224	211252.670	47.348	182	20	7.7	3120.000		20.000
320	539811.094	211258.839	47.587	182	35	24.5	3130.000		29.999
325	539851.249	211288.624	48.756	183	51	48.1	3180.000	79.983	10.000
326	539859.437	211294.365	48.984	184	7	4.8	3190.000	89.976	10.000
327	539867.675	211300.033	49.211	184	22	21.6	3200.000	99.967	10.000

POINT 317 TO POINT 0306
BEARING 228 48 24.4
DISTANCE 99.989

Explanatory notes

1. The headings are reasonably self-explanatory. X and Y are the horizontal co-ordinates and Z is the reduced level of the centre line of the road.

2. This section of road commences with a straight length from point 285 to 305, a distance of just under 200 m, and then goes into a transition curve (indicated) followed by a circular curve (not shown).

3. Station STB9 is about 15 m offset from the centre-line, probably just inside the fence line.

Bibliography

Except as noted, the following published sources were in print as at April 1987. Where not immediately apparent from the titles, the sources are recommended for information on the topic(s) indicated.

BANNISTER, A and RAYMOND, S. *Surveying*, sixth edition Longman, London. 1986
  - curve ranging
  - theodolite traverses
  - tacheometry

BRIGHTY, S.G. *Setting out: a guide for site engineers* Granada, London. 1981
  - general use of survey instruments
  - setting-out roads

BRITISH STANDARDS INSTITUTION *Radiation safety of laser products and systems* BS 4803: Part 1: 1983, General and Part 3: 1983, *Guidance for users*

BRITISH STANDARDS INSTITUTION *Code of practice for accuracy in building* BS 5606: 1978

BUILDING RESEARCH ESTABLISHMENT *Accuracy in setting-out* BRE Digest 234. HMSO, London. 1980
  - achievable accuracy

CHENEY. J.E. *Techniques and equipment using the surveyor's level for accurate measurement of building movement* Proc. BGS Symposium on field instrumentation, Paper 2, pp 85-89, Butterworth, London. 1973 (Also as BRE Current Paper 26/73, out of print but amended photocopy available from author at BRE, pending re-publication)
  - precise levelling
  - details of BRE levelling station

CONSTRUCTION INDUSTRY RESEARCH AND INFORMATION ASSOCIATION *Medical code of practice for work in compressed air* Report 44, Third Edition, London. 1982

COUNTY SURVEYORS' SOCIETY *Highway transition curve tables (metric)* Carriers Publishers Co. London, 1969 (Available from Drydens (Printers) Ltd, 192 Brent Crescent, London NW10 7XU)

COX, E.A. *Safety in the use of lasers on site* CIOB Technical Information Paper 22, 1983

DEPARTMENT OF TRANSPORT *Specification for highway works* HMSO, LONDON. 1986

HARLEY, J.B. *Ordnance Survey maps–a descriptive manual* Ordnance Survey, Southampton. 1975

IRVINE, D.J. and SMITH, R.J.H. *Trenching practice* CIRIA Report 97. London, 1983

NATIONAL JOINT HEALTH AND SAFETY COMMITTEE FOR THE WATER SERVICE *Safe working in sewers and at sewage works* Health and Safety Guidline No 2. Water Authorities Association, London. 1979

NATIONAL WATER COUNCIL: BRITISH GAS CORPORATION *Model consultative procedures for pipe construction involving deep excavation* January 1983

ROYAL INSTITUTION OF CHARTED SURVEYORS *A guide to the safe use of lasers in surveying and construction* London, 1980 pp 9.1 to 9.3 (No longer in print)

SCHOFIELD, W. *Engineering Surveying*, Volume 1 Third edition, Butterworths, London. 1984
- theodolite traverses
- use of gyro-theodolite

Ibid, **Vol 2**, Second edition, London. 1984
- electronic distance-measurement (EDM)

SHEPHERD, F.A. *Engineering surveying—problems and solutions* Second edition. Edward Arnold, London. 1983
- survey formulae and sample calculations

SWEDISH INSTITUTE FOR BUILDING RESEARCH *Measuring practice on the building site* Bulletin M83:16, CIB/FIG Report 69

UREN, J., AND PRICE, W.F. *Calculations for engineering surveying* Van Nostrand Reinhold (UK), Wokingham. 1984
- survey formulae and sample calculations

SUGGESTED COLOUR CODE FOR ROAD PEGS AND PROFILES

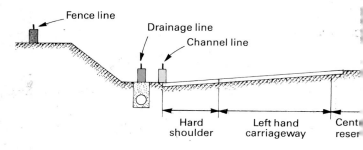

MOTORWAY OR DUAL CARRIAGEWAY (schematic)

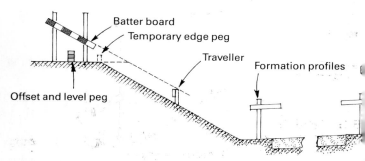

LEFT HAND OR SINGLE CARRIAGEWAY

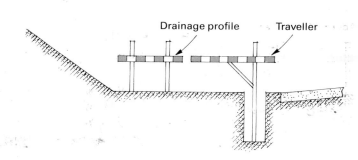

DUAL OR SINGLE CARRIAGEWAY